SpringerBriefs in Applied Sciences and Technology

Nanoscience and Nanotechnology

Series editors

Hilmi Volkan Demir, Nanyang Technological University, Singapore, Singapore
Alexander O. Govorov, Ohio University, Athens, USA

Nanoscience and nanotechnology offer means to assemble and study superstructures, composed of nanocomponents such as nanocrystals and biomolecules, exhibiting interesting unique properties. Also, nanoscience and nanotechnology enable ways to make and explore design-based artificial structures that do not exist in nature such as metamaterials and metasurfaces. Furthermore, nanoscience and nanotechnology allow us to make and understand tightly confined quasi-zero-dimensional to two-dimensional quantum structures such as nanoplatelets and graphene with unique electronic structures. For example, today by using a biomolecular linker, one can assemble crystalline nanoparticles and nanowires into complex surfaces or composite structures with new electronic and optical properties. The unique properties of these superstructures result from the chemical composition and physical arrangement of such nanocomponents (e.g., semiconductor nanocrystals, metal nanoparticles, and biomolecules). Interactions between these elements (donor and acceptor) may further enhance such properties of the resulting hybrid superstructures. One of the important mechanisms is excitonics (enabled through energy transfer of exciton-exciton coupling) and another one is plasmonics (enabled by plasmon-exciton coupling). Also, in such nanoengineered structures, the light-material interactions at the nanoscale can be modified and enhanced, giving rise to nanophotonic effects.

These emerging topics of energy transfer, plasmonics, metastructuring and the like have now reached a level of wide-scale use and popularity that they are no longer the topics of a specialist, but now span the interests of all "end-users" of the new findings in these topics including those parties in biology, medicine, materials science and engineerings. Many technical books and reports have been published on individual topics in the specialized fields, and the existing literature have been typically written in a specialized manner for those in the field of interest (e.g., for only the physicists, only the chemists, etc.). However, currently there is no brief series available, which covers these topics in a way uniting all fields of interest including physics, chemistry, material science, biology, medicine, engineering, and the others.

The proposed new series in "Nanoscience and Nanotechnology" uniquely supports this cross-sectional platform spanning all of these fields. The proposed briefs series is intended to target a diverse readership and to serve as an important reference for both the specialized and general audience. This is not possible to achieve under the series of an engineering field (for example, electrical engineering) or under the series of a technical field (for example, physics and applied physics), which would have been very intimidating for biologists, medical doctors, materials scientists, etc.

The Briefs in NANOSCIENCE AND NANOTECHNOLOGY thus offers a great potential by itself, which will be interesting both for the specialists and the non-specialists.

More information about this series at http://www.springer.com/series/11713

Talha Erdem · Hilmi Volkan Demir

Color Science and Photometry for Lighting with LEDs and Semiconductor Nanocrystals

 Springer

Talha Erdem
Cavendish Laboratory
University of Cambridge
Cambridge, UK

Hilmi Volkan Demir
School of Electrical and Electronic
Engineering, School of Physical
and Mathematical Sciences, and School
of Materials Science and Engineering
Nanyang Technological University
Singapore, Singapore

and

Institute of Materials Science
and Nanotechnology (UNAM)
Bilkent University
Çankaya, Ankara, Turkey

ISSN 2191-530X   ISSN 2191-5318   (electronic)
SpringerBriefs in Applied Sciences and Technology
ISSN 2196-1670   ISSN 2196-1689   (electronic)
Nanoscience and Nanotechnology
ISBN 978-981-13-5885-2   ISBN 978-981-13-5886-9   (eBook)
https://doi.org/10.1007/978-981-13-5886-9

Library of Congress Control Number: 2018966833

This Springer imprint is published by the registered company Springer Nature Singapore Pte Ltd.
The registered company address is: 152 Beach Road, #21-01/04 Gateway East, Singapore 189721, Singapore

Contents

Chapter 1
Introduction

Abstract Here we briefly emphasize the importance of lighting for our daily lives as well as its role in energy consumption. We very briefly introduce the problems that need to be addressed and finally summarize the contents of this brief.

Keywords Lighting · Energy consumption · LEDs

Light is an essential part of the human life and is considered an important trigger for the development of culture and knowledge. In modern times, light and together with it light-emitting devices including lamps, lasers, and displays have become an inseparable part of our lifestyle. Acknowledging this importance of light and underlying scientific breakthroughs, UNESCO announced 2015 as the "International Year of Light and Light-based Technologies" [1].

The significance of light shows itself in its share within the total energy consumption. Decreasing this amount is expected to substantially contribute to the efforts of mitigating the carbon footprint; therefore, there is a strong demand for developing efficient light sources [2]. Research addressing this need has already started to help reduce the share of the energy consumed by the lighting from ~20% in 2007 [3] to 15% in 2015 [4]. The driving force for this development has been the transition from the traditional light sources to the light-emitting diodes (LEDs) [5]. As tabulated by the US Department of Energy [6], an LED-based lamp consumes only ca. 20% of the energy that an incandescent lamp typically uses to deliver a similar brightness level. The US Department of Energy predicts that by 2030 the transition to LEDs will enable a total of ~40% energy saving. In addition to this saving, the bulb lifetime, which is 1000 h for incandescent lamps reaches, 25,000 h for the LED based lamps. This is also an important advantage of using LEDs to decrease the cost [6].

Two main strategies are followed to realize white-light emission using LEDs. The most straightforward approach is the collective use of multiple LED chips each individually emitting in different colors. However, despite being straightforward, this method of producing white light is significantly costly due to the driving electrical circuitry. In addition, different material systems required for such LEDs of varying color components further increases the production complexity and cost. More importantly, the efficiencies of the green and yellow LED chips are commonly low;

© The Author(s), under exclusive licence to Springer Nature Singapore Pte Ltd. 2019
T. Erdem and H. V. Demir, *Color Science and Photometry for Lighting with LEDs and Semiconductor Nanocrystals*, Nanoscience and Nanotechnology,
https://doi.org/10.1007/978-981-13-5886-9_1

therefore, the white LED luminaries using these LED chips suffer from low efficiencies. As a consequence, multi-chip approach for white light generation has not been able to find ubiquitous use. A more common method for this purpose relies on the hybridization of color converters with LED chips. In this method, a blue or near-ultraviolet (UV) LED excites the color converting material that is coated on top of the LED chip. Currently, the most common color converters are the phosphors made of rare-earth ions. These phosphors possessing near unity quantum efficiencies are typically very broad emitters spanning the spectral range from 500 to 700 nm. This spectral broadness allowing for white light generation is, however, their plague because the emission spectra of the phosphors extend toward the spectral region where the human eye is not sensitive anymore. It is also very difficult to fine-tune the spectrum of the LEDs using phosphors to increase the color quality by increasing the color rendering capability and shade of the white light [3, 7]. Another problem associated with these phosphors is the supply problems of the rare-earth elements threatening their future in optoelectronics [8]. At this point, narrow-band emitters such as colloidal nanocrystal quantum dots step forward as they enable spectral fine-tuning [9] while the saturated colors emitted by them allows for obtaining displays that can define colors as opposed to broad-emitters such as phosphors [10–12].

While designing light sources made of narrow-band emitters, one of the most important questions is how to achieve high quality and high efficiency. In this brief, we aim to establish guidelines to answer this question for indoor, outdoor, and display lighting applications. We start with the technical background on light stimulus and human eye, then continue with colorimetry and photometry. Next, we describe the guidelines for designing light sources made of narrow-band emitters in the order of indoor lighting, outdoor lighting, and display backlighting. Finally, we conclude this brief with a future perspective.

References

1. UNESCO (2014) The International Year of Light
2. Phillips JM et al (2007) Research challenges to ultra-efficient inorganic solid-state lighting. Laser Photonics Rev 1(4):307
3. Krames MR et al (2007) Status and future of high-power light-emitting diodes for solid-state lighting. J Disp Technol 3(2):160–175
4. US Department of Energy "How much electricity is used for lighting in the United States?" [Online]. Available: https://www.eia.gov/tools/faqs/faq.cfm?id=99&t=3. Accessed 14 Jun 2010
5. US Department of Energy (2014) "Energy savings forecast of solid-state lighting in general illumination applications
6. US Department of Energy "How energy-efficient light bulbs compare with traditional incandescent." [Online]. Available: http://energy.gov/energysaver/how-energy-efficient-light-bulbs-compare-traditional-incandescents. Accessed 14 Jun 2016
7. Müller-Mach R, Müller GO, Krames MR, Trottier T (2002) High-power phosphor-converted light-emitting diodes based on III-Nitrides. IEEE J Sel Top Quantum Electron 8(2):339
8. Graydon O (2011) The new oil? Nat Photonics 5(1):1

9. Erdem T, Demir HV (2011) Semiconductor nanocrystals as rare-earth alternatives. Nat Photonics 5(1):126
10. Erdem T, Demir HV (2013) Color science of nanocrystal quantum dots for lighting and displays. Nanophotonics 2(1):57–81
11. Jang E, Jun S, Jang H, Lim J, Kim B, Kim Y (2010) White-light-emitting diodes with quantum dot color converters for display backlights. Adv Mater 22(28):3076–3080
12. Luo Z, Chen Y, Wu S-T (2013) Wide color gamut LCD with a quantum dot backlight. Opt Express 21(22):26269–26284

Chapter 2
Light Stimulus and Human Eye

Abstract In this Chapter, we summarize the structure of the human eye and introduce the sensitivity functions of various photoreceptors and present the visual regimes and corresponding eye sensitivity functions.

Keywords Human eye · Eye sensitivity function · Colorimetry · Photometry

In order to design high quality and highly efficient light sources, quantitative measures of light stimuli are necessary. In its broadest sense, we can classify the light stimulation in several categories that are actinometry, radiometry, photometry, and colorimetry [1]. Among these, actinometry and radiometry only consider the physical nature of light while photometry and colorimetry take the interaction of light with the human visual system into account.

Actinometry is interested in the particle nature of light and works with light quanta, i.e., photons. According to this classification, the amount of light is expressed in number of photons. Based on this, the light amount per unit time is expressed in number of photons per second, the amount of light per unit time per unit area is given in number of photons per second per meter square, etc. Radiometry, on the other hand, only deals with the wave nature of light rather than its particle behavior. It employs energy to express the amount of light, usually in the units of Joules. Then, the amount of light per unit time becomes actually the power of light and typically expressed in Watts. The irradiance, i.e., the amount of light per unit time per unit area is the power per unit area. Another important quantity in radiometry is the radiance which stands for the amount of light per unit area per unit solid angle.

Both colorimetry and photometry unify the human perception of light with its physical nature. Therefore, both of these categories are strongly related to actinometry and especially to radiometry. Photometry aims to quantify the visual effectiveness of light by considering all the elements of human visual system as a single body. It expresses the amount of light in lumens, which basically stands for the perceived optical power and all the light related quantities are based on this unit. Different than photometry, colorimetry evaluates the light stimuli based on the color perception. It focuses on quantifying the perceived color of an arbitrary light stimulus. From this perspective, it is significantly different than the others and the quantities that

T. Erdem and H. V. Demir, *Color Science and Photometry for Lighting with LEDs and Semiconductor Nanocrystals*, Nanoscience and Nanotechnology, https://doi.org/10.1007/978-981-13-5886-9_2

it defines are closely related to photoreceptors in the human eye. Therefore, before going into the details of colorimetry, we find it beneficial to present a brief review of the properties of human eye.

2.1   Human Eye

We perceive the world through our eyes. To understand the process of vision, knowing how we see is essential, especially for the colorimetric quantification of light. As shown in Fig. 2.1, cornea is the transparent layer of the eye where the light first enters. After passing the anterior chamber, light reaches the lens that focuses the light on the retina, which is full of visual neurons transmitting the visual information to brain [2].

The visual neurons have three main layers [3]: photoreceptors, intermediate neurons, and ganglion cells. The latter two layers mainly serve as signal carriers to brain whereas the photoreceptors are the light-sensitive cells in our eyes. They have three types, which are rods, cones, and melanopsin.

Rods and cones, which are named in accordance with their shape, are responsible for the visual perception. Rods that sense the whole visible color regime without any color differentiation are found more ubiquitously on the retina compared to cones. Due to their lack of color differentiation, these cells do not contribute to color perception; however, they are much more sensitive at low light levels. Cones, on the other hand, have three types known as S-cones, M-cones, and L-cones standing for short, medium, and long wavelength sensitive photoreceptors. In other words, S-cones are responsible for the blue color perception while the M- and L-cones perceive green-yellow and red colors (Fig. 2.2). In addition to their differences in color perception, the cones and the rods also differ in their activity levels at different light levels. Under dim lighting conditions, the rods govern the visual process while cones do not have a meaningful contribution. As a result, we cannot see the colors of

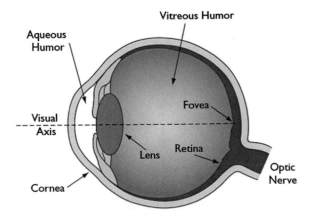

Fig. 2.1 Schematics of the human eye. Reproduced with permission from Ref. [1]. © Optical Society of America 2003

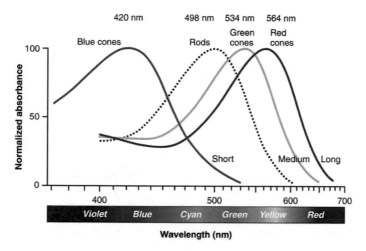

Fig. 2.2 Sensitivity spectra of rod and cone photoreceptors. Adapted from Ref. [4]

the objects in dark. On the other hand, at high ambient lighting levels the contribution of rods to the vision remains limited while we see the world via the signals received by cones. This enables us to perceive the world in color. The dark-adapted visual regime, in which the vision is governed by the rods, is called the scotopic regime. The regime of high light levels where the cones dominate the visual perception is referred to as the photopic vision regime. The scotopic regime corresponds to dark conditions while room lighting or brighter environments fall into the category of photopic regime. At mediocre lighting levels such as street lighting, both the rods and the cones contribute to the visual perception together, this level of lighting conditions is known as the mesopic regime. Our visual perception adapts to the ambient lighting levels by setting the contribution of rods and cones in vision. As a result, the average sensitivity of our eyes depends strongly on the visual regime (Fig. 2.3) that our eyes are subject to. This brings about the necessity of quantifying photometric parameters in accordance with the ambient lighting levels which we will discuss in the next sections in detail.

The third photoreceptor melanopsin, which was discovered in early 2000s, does not play a significant role in vision. Nevertheless, it has a crucial role in the regulation of the circadian cycle, i.e., the daily biological rhythm [5, 6]. Melanopsin plays this role by controlling the secretion of the melatonin hormone whose concentration is a signal to the body for the time of the day. During the daytime, melatonin secretion is suppressed, and brain interprets this decrease as the daytime signal while the brain interprets the increase of the melatonin concentration as the night time signal. Although it is currently well known that the lighting affects the secretion of melatonin contributing to the control of the biological rhythm, it is still controversial how the suppression of melatonin occurs and how lighting affects it. According to Rea, melatonin suppression is affected collectively by the rods, cones, and melanopsin [7], while Gall [8] and Enezi et al. [9] employ a simpler model and connects the melatonin

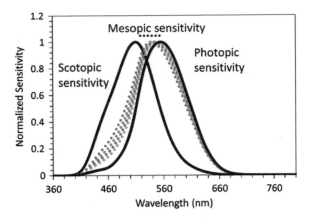

Fig. 2.3 Average eye sensitivity functions as a function of the light levels

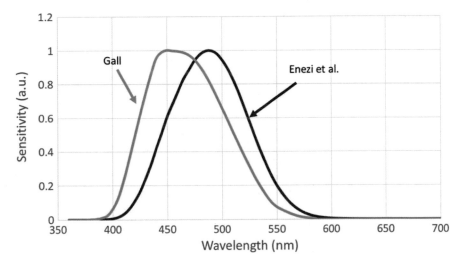

Fig. 2.4 Relative sensitivity spectra of melatonin suppression according to Gall [8] and Enezi et al.
[9]

suppression only to the effect of lighting on the melanopsin since some neurons in the
brain robustly react to the melanopsin activity but not to that of the cones [10]. The
models developed by Gall and Enezi predict different circadian sensitivity spectra
(Fig. 2.4). The striking feature of both spectra is that they cover the blue part of
the visible regime. Actually, we may also qualitatively guess the contribution of the
blue range by looking at the sun's spectrum at different times of the day. Before the
noon and during the afternoon hours, blue content in the sun's spectrum is much
stronger than the red-shifted spectrum in the evening times. As an adaptation to this
variation, our body reacts to the blue content of the sun by reducing the melatonin
concentration to send the daytime signal and vice versa. Related to this phenomenon,

the duration of exposure to natural and/or artificial light sources affects the circadian rhythm and consequently the human health. For example, insufficient exposure to bluish light in the morning shifts the circadian cycle [11] and artificial lighting with strong blue content leads to the melatonin suppression. This means that great care should be taken while designing artificial lighting and displays.

References

1. Packer O, Williams DR (2003) Light, the retinal image, and photoreceptors. In: Shevell SK (ed) The science of color: second edition, Elsevier Science and Technology, pp 41–102
2. Wyszecki G, Stiles WS (1983) Color science: concepts and methods, quantitative data and formulae, 2nd edn, vol. 8, no. 4. Wiley, New York
3. Stell WK (1972) The morphological organization of the Vertebrate Retina. In: Fuortes M (ed) Physiology of photoreceptor organs. Handbook of sensory physiology Springer, Berlin, Heidelberg
4. Open Stax Courses "Sensory perception." In: Anatomy & physiology. Online at: https://cnx.org/contents/FPtK1zmh@6.27:s3XqfSLV@4/Sensory-Perception. Accessed 12 Nov 2018
5. Hattar S, Liao HW, Takao M, Berson DM, Yau KW (2002) Melanopsin-containing retinal ganglion cells: architecture, projections, and intrinsic photosensitivity. Science (80-) 295(5557):1065–1070
6. Berson DM, Dunn FA, Takao M (2002) Phototransduction by retinal ganglion cells that set the circadian clock. Science 295(557):1070–1073
7. Rea MS, Figueiro MG, Bierman A, Bullough JD (2010) Circadian light. J Circadian Rhythms 8(1):2
8. Gall D (2004) Die Messung circadianer Strahlungsgrößen In: Proceedings of 3. Internationales Forum fur den lichttechnischen Nachswuchs
9. Enezi JA, Revell V, Brown T, Wynne J, Schlangen L, Lucas R (2011) A "melanopic" spectral efficiency function predicts the sensitivity of melanopsin photoreceptors to polychromatic lights. J Biol Rhythms 26(4):314–323
10. Brown TM, Wynne J, Piggins HD, Lucas RJ (2011) Multiple hypothalamic cell populations encoding distinct visual information. J Physiol 589(Pt 5):1173–1194
11. Figueiro MG, Rea MS (2010) Lack of short-wavelength light during the school day delays dim light melatonin onset (DLMO) in middle school students. Neuroendocrinol Lett 31(1):92–96

Chapter 3
Colorimetry for LED Lighting

Abstract In this Chapter, we explain the basics of colorimetry and introduce the colorimetric tools useful for designing light sources.

Keywords Colorimetry · Color matching · Chromaticity diagram

For general lighting, a good white light source should help us perceive the real colors of objects as accurately as possible. Especially from the architectural point of view, we also need to be able to compare the white light emitted by different light sources. From the point of displays, the light sources should be able to reproduce the colors of objects as correctly as possible. To evaluate all these qualities of light sources, colorimetry plays an essential role. It provides us with a quantitative description of colors and gives us the tools kit to test the quality of a light source for various applications and compare different light sources.

The human color perception forms the basis of the colorimetry field. As we have discussed in the previous section, the cones are the photoreceptors that form the essence of our color perception. When we look at their sensitivity spectra (Fig. 2.2), we observe that three types of cones predominantly absorb across different parts of the visual spectrum. This is the reason why we perceive three primary colors. However, another important feature of these sensitivity spectra is that they also overlap very strongly. This means—from a mathematical point of view—they do not form an orthogonal basis. As a result, we may perceive different combinations of the light stimuli having different spectra as identically the same color. This enables to achieve perfect or at least satisfactory perceived color accuracy without mimicking the sun's spectrum.

The attempts to quantitatively describe the colors date back to the early 20th century with the color wheel and color triangle of J. C. Maxwell. There have been several additional efforts on this topic and in 1931, the International Commission on Illumination (CIE) introduced a standard quantitative description of color by mapping the perceived colors to a color space called CIE 1931. This color space makes use of three color matching functions: $\bar{x}$, $\bar{y}$, and $\bar{z}$, whose spectral distributions are given in Fig. 3.1 [1].

T. Erdem and H. V. Demir, *Color Science and Photometry for Lighting with LEDs and Semiconductor Nanocrystals*, Nanoscience and Nanotechnology, https://doi.org/10.1007/978-981-13-5886-9_3

11

Fig. 3.1 Spectral distribution of the color matching functions used in CIE 1931 color space

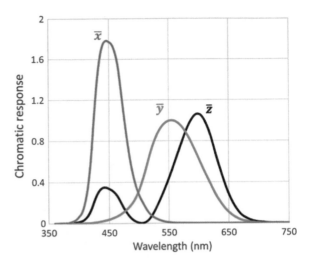

In order to compute the color coordinates, we first calculate the so-called tristimulus values, X, Y, and Z by using Eqs. (3.1)–(3.3) for an arbitrary radiation spectrum of $s(\lambda)$.

$$X = \int s(\lambda)\bar{x}(\lambda)d\lambda \tag{3.1}$$

$$Y = \int s(\lambda)\bar{y}(\lambda)d\lambda \tag{3.2}$$

$$Z = \int s(\lambda)\bar{z}(\lambda)d\lambda \tag{3.3}$$

The (x, y) chromaticity coordinates, which are also referred to as CIE 1931 chromaticity coordinates, are calculated using Eqs. (3.4)–(3.6). Instead of using three independent variables, the normalization reduces the coordinates to (x, y) as $z = 1 - x - y$. Since one of the three coordinates is dependent on the other remaining two, this methodology generates a two-dimensional color mapping as presented in Fig. 3.2.

$$x = \frac{X}{X + Y + Z} \tag{3.4}$$

$$y = \frac{Y}{X + Y + Z} \tag{3.5}$$

$$z = \frac{Z}{X + Y + Z} = 1 - x - y \tag{3.6}$$

Despite the fact that this color mapping is the most widely preferred chromaticity diagram, it has an inherent problem that the geometrical difference between the posi-

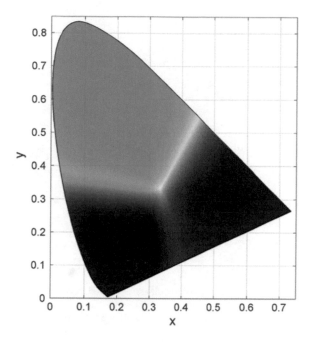

Fig. 3.2 (x, y) chromaticity diagram. This color gamut is also known as CIE 1931 chromaticity diagram

tions of pairs of colors does not consistently correspond to the perceived difference between the colors leading to nonuniform color distributions. As a solution to this problem, additional color mapping methodologies were proposed by CIE. Among them are the (u, v), (u', v'), and $L*a*b*$ chromaticity diagrams.

The (u, v) and (u', v') coordinates are related to X, Y, and Z color coordinates using Eqs. (3.7)–(3.9). We present the (u', v') chromaticity diagram in Fig. 3.3. As we can clearly see, especially green and red colors are more equally distributed on this diagram.

$$u = u' = \frac{4X}{X + 15Y + 3Z} \tag{3.7}$$

$$v = \frac{6Y}{X + 15Y + 3Z} \tag{3.8}$$

$$v' = \frac{3}{2}v \tag{3.9}$$

Despite the improvements on (u', v') chromaticity diagrams in terms of color uniformity, this system still needed to be improved. In addition to this, the existing systems, which do not include the effect of the luminance on the color perception, needed to be modified to possess this information. These issues were addressed by CIE in 1976 and $(L^*a^*b^*)$ chromaticity diagram was introduced (Fig. 3.4). Contrary to the previous systems, $(L^*a^*b^*)$ is a three-dimensional color space that maps the perceived colors considering the effects of luminance. The corresponding color

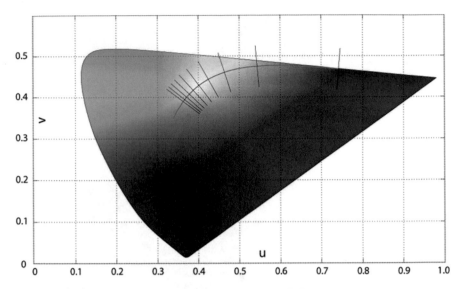

Fig. 3.3 (u′, v′) chromaticity diagram. Reproduced from Ref. [2]

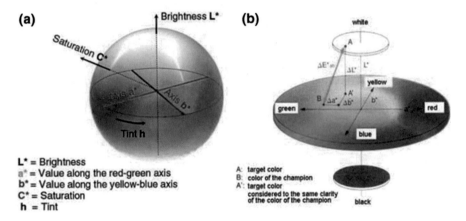

L* = Brightness
a* = Value along the red-green axis
b* = Value along the yellow-blue axis
C* = Saturation
h = Tint

A: target color
B: color of the champion
A′: target color
    considered to the same clarity
    of the color of the champion

Fig. 3.4 Illustration of (**a**) the full CIE L*a*b* chromaticity diagram and (**b**) a cross-section. Reproduced with permission from Ref. [3]

coordinates are calculated using Eqs. (3.10)–(3.12), where X_n, Y_n, and Z_n are the nominally white object color stimulus, and calculated mostly using CIE standard illuminant A.

$$L^* = 116\left(\frac{Y}{Y_n}\right)^{1/3} \tag{3.10}$$

$$a^* = 500\left[\left(\frac{X}{X_n}\right)^{1/3} - \left(\frac{Y}{Y_n}\right)^{1/3}\right] \tag{3.11}$$

$$b^* = 200 \left[\left(\frac{X}{X_n} \right)^{1/3} - \left(\frac{Z}{Z_n} \right)^{1/3} \right] \tag{3.12}$$

To express color differences under different conditions, color adaptation transformations are developed. This approach allows to a quantitative description of human color perception adaptation to different white point/white light changes under different illumination conditions. Here, we will summarize CMCCAT2000 method, which is essentially a developed version of the previous adaptation transformation CIECAT97. This transformation makes use of X, Y and Z values of a spectral power distribution usually from the reflected stimulus of the test source (called X_s, Y_s and Z_s), the chromaticity coordinates of the spectral power distribution of the test light source (dubbed with X_t, Y_t and Z_t), the chromaticity coordinates of the spectral power distribution of a reference light source such as standard D65 illuminant (dubbed as X_r, Y_r and Z_r), and finally the luminance values of test and reference adapting fields, named as L_{a1} and L_{a2}. The calculation starts with transforming X, Y, and Z tristimulus values of all input tristimulus values to R, G, and B values using the relation given below:

$$\begin{bmatrix} R \\ G \\ B \end{bmatrix} = \begin{bmatrix} 0.7982 & 0.3389 & -0.1371 \\ -0.5918 & 1.5512 & 0.0406 \\ 0.0008 & 0.0239 & 0.9753 \end{bmatrix} \begin{bmatrix} X \\ Y \\ Z \end{bmatrix} \tag{3.13}$$

Next, the degree of adaptation D is 1 if $D' > 1$, and it is 0 if $D' < 0$, and otherwise it is equal to D' where D' is found using Eq. (3.14).

$$D' = 0.08 \log_{10}(0.5(L_{a1} + L_{a2})) + 0.76 - \frac{0.45(L_{a1} - L_{a2})}{L_{a1} + L_{a2}} \tag{3.14}$$

The adapted RGB values (R_c, G_c and B_c) are calculated using the relation given below in Eq. (3.15):

$$\begin{bmatrix} R_c \\ G_c \\ B_c \end{bmatrix} = \begin{bmatrix} D \times \frac{R_r}{R_t} + 1 - D & 0 & 0 \\ 0 & D \times \frac{G_r}{G_t} + 1 - D & 0 \\ 0 & 0 & D \times \frac{B_r}{B_t} + 1 - D \end{bmatrix} \begin{bmatrix} R_s \\ G_s \\ B_s \end{bmatrix} \tag{3.15}$$

The adapted X, Y, Z tristimulus values (X_c, Y_c and Z_c) are found using Eq. (3.16).

$$\begin{bmatrix} X_c \\ Y_c \\ Z_c \end{bmatrix} = \begin{bmatrix} 0.7982 & 0.3389 & -0.1371 \\ -0.5918 & 1.5512 & 0.0406 \\ 0.0008 & 0.0239 & 0.9753 \end{bmatrix}^{-1} \begin{bmatrix} R_c \\ G_c \\ Z_c \end{bmatrix} \tag{3.16}$$

Finally, the adapted x, y and z chromaticity coordinates are computed using Eqs. (3.4)–(3.6).

All of these calculations are necessary for evaluating the color rendition performance of the sources and the shade of the light. In Appendix A of this brief, we provide MATLAB codes for calculating the presented chromaticity coordinates along with color matching function tables.

References

1. "CIE Commission Proceedings," 1931
2. "By Adoniscik—Own work, CC BY 3.0, https://commons.wikimedia.org/w/index.php?curid=3838965
3. Lorusso S, Natali A, Matteucci C (2007) Colorimetry applied to the field of cultural heritage: examples of study cases. Conservation Sci Cultural Heritage 7:187–220

Chapter 4
Metrics for Light Source Design

Abstract In this part of this brief, we summarize the metrics that need to be considered for designing light sources. We start with metrics on the shade of color and then continue with color rendering and photometry.

Keywords Color temperature · Color rendering index · Color quality scale · Luminous efficiency

4.1 Cool Versus Warm White Light: Correlated Color Temperature (CCT)

The chromaticity diagrams offering color uniformity are especially targeted for comparing the colors of different sources. For a white light source, one of the obvious illuminants whose color is compared with is the sun. Since the sun is a blackbody radiator, the shade of the white light radiated by the designed light source can be safely compared with the shade of a blackbody radiator whose spectral distribution $P(\lambda)$ is given below.

$$P(\lambda) = \frac{2\pi c^2 h}{\lambda^5} \frac{1}{e^{hc/(\lambda kT)} - 1} \qquad (4.1)$$

where c is the speed of light, h is the Planck's constant, k is the Boltzmann constant, and T is the temperature.

The emission spectrum of a blackbody radiator is a function of its temperature. With the same analogy, the shade of the white light of an arbitrary white light source can be characterized by finding the temperature of the blackbody radiator whose color is closest to the color of the light source. This temperature is called the correlated color temperature (CCT). As opposed to the common usage in thermodynamics, high CCTs indicate a cool white-shade since a blackbody radiator at higher temperatures have a stronger bluish color tint. Similarly, a blackbody radiator at lower temperatures have a stronger red content giving its emission a warmer white shade

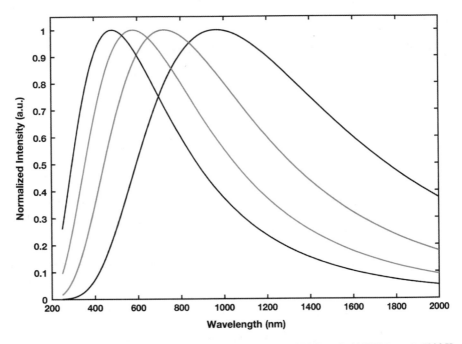

Fig. 4.1 Spectral power distribution of blackbody radiators at 3000 K (red), 4000 K (green), 5000 K (blue), and 6000 K (violet)

(Fig. 4.1). Traditionally, the CCT of an arbitrary light source is calculated using (u', v') chromaticity diagram (see Fig. 3.3). Incandescent light bulbs have CCTs around 3000 K and fluorescent tubes have varying CCTs from 3000 to 6500 K, whereas the CCT of the sun is close to 6000 K [1]. Having a warmer white shade (between 3000 and 4500 K) is more desirable for indoor lighting applications mainly for avoiding the disturbing effects of cool white light on the human biological clock. In the Appendix B of this brief we provide codes for calculating the correlated color temperature of a given spectral power distribution.

4.2 Color Rendition: Color Rendering Index (CRI), Color Quality Scale (CQS), and Other Metrics

A critical parameter regarding the performance of a light source is its capability to render the real colors of the objects. When objects are illuminated with a high-quality light source, we expect to perceive the colors correctly. This requirement has to be addressed especially for the indoor lighting applications. Moreover, for outdoor lighting applications such as road lighting, a light source with good color rendering

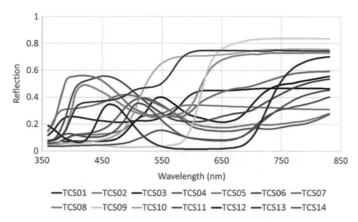

Fig. 4.2 Reflection spectra of the test color samples (TCS) used for calculating the color rendering index

capability was shown to increase the safety of roads and streets for pedestrians and drivers as good color rendition helps increase the color contrast [2].

This property of light sources has been proposed to be evaluated by various measures including the color discrimination index [3], color rendering capacity [4], feeling of contrast index [5], and flattery index [6]. However, these metrics have not attracted considerable attention in the lighting community to date. Therefore, we will not cover them here in detail and continue with two of the most commonly used color rendition metrics, which are the color rendering index (CRI) and the color quality scale (CQS) [7].

CRI was first introduced by CIE in 1971 [8] and later in 1995 its calculation method was revised [9]. It makes use of fourteen test samples whose reflection spectra are given in Fig. 4.2 and the table summarizing these spectra are given in Appendix A. The calculation assumes that the reference white light source, which is in general a blackbody radiator, renders the colors of objects perfectly. The calculation involves evaluating the performance of the test light sources by comparing reflection spectra of the reference and test light sources from the test color samples and calculating the associated color difference between these two light sources. This color difference data was then employed to calculate the CRI whose maximum value is 100 indicating a perfect color rendition capability. Its minimum value is -100 which indicates the worst color rendition performance. During the CRI calculation, a color rendering index value specific to each test sample is obtained. The general color rendering index is calculated by using the first eight test samples while the remaining six samples define the specific CRI. In general, a light source possessing CRI > 90 is considered to successfully render the real colors of objects [10].

Calculation of CRI starts with the determination of (u, v) coordinates of the reflection from the test sample i using the reference (dubbed with ref) and test light sources. Using Eqs. (4.2) and (4.3), (u, v) coordinates are transformed to (c, d) coordinates.

$$c = \frac{4 - u - 10v}{v} \tag{4.2}$$

$$d = \frac{1.708v + 0.404 - 1.481u}{v} \tag{4.3}$$

Subsequently, $(u^{**}_{text,i}, v^{**}_{text,i})$ coordinates are found using Eqs. (4.4) and (4.5).

$$u^{**}_{test,i} = \frac{10.872 + 0.404 \frac{c_{ref}}{c_{test}} c_{test,i} - \frac{4d_{ref}}{d_{test}} d_{test,i}}{16.518 + 1.481 \frac{c_{ref}}{c_{test}} c_{test,i} - \frac{d_{ref}}{d_{test}} d_{test,i}} \tag{4.4}$$

$$v^{**}_{test,i} = \frac{5.520}{16.518 + 1.481 \frac{c_{ref}}{c_{test}} c_{test,i} - \frac{d_{ref}}{d_{test}} d_{test,i}} \tag{4.5}$$

Then, $(u^{**}_{text}, v^{**}_{text})$ are obtained using Eqs. (4.6) and (4.7).

$$u^{**}_{test,i} = \frac{10.872 + 0.404c_{ref} - 4d_{ref}}{16.518 + 1.481c_{ref} - d_{ref}} \tag{4.6}$$

$$u^{**}_{test,i} = \frac{5.520}{16.518 + 1.481c_{ref} - d_{ref}} \tag{4.7}$$

The color shifts for each test sample (ΔE^{**}_i) are calculated with Eqs. (4.8)–(4.11)

$$\Delta L^{**} = \left(25Y^{\frac{1}{3}}_{ref,i} - 17\right) - \left(25Y^{\frac{1}{3}}_{test,i} - 17\right) = L^{**}_{ref,i} - L^{**}_{test,i} \tag{4.8}$$

$$\Delta u^{**} = 13L^{**}_{ref,i}\left(u_{ref,i} - u_{ref}\right) - 13L^{**}_{test,i}\left(u_{test,i} - u_{test}\right) \tag{4.9}$$

$$\Delta v^{**} = 13L^{**}_{ref,i}\left(v_{ref,i} - v_{ref}\right) - 13L^{**}_{test,i}\left(v_{test,i} - v_{test}\right) \tag{4.10}$$

$$\Delta E^{**}_i = \sqrt{(\Delta L^{**})^2 + (\Delta u^{**})^2 + (\Delta v^{**})^2} \tag{4.11}$$

Following the computation of the color shift, CRI for each test sample is calculated using Eq. (4.12). Finally, the general CRI can be found using Eq. (4.13).

$$CRI_i = 100 - 4.6\Delta E^*_i \tag{4.12}$$

$$CRI = \frac{1}{8} \sum_{i=1}^{8} CRI_i \tag{4.13}$$

In Appendix B of this brief, we also provide MATLAB codes for calculating the CRI for a given spectral power distribution.

Although CRI still remains as the most frequently used measure of color rendition, it suffers from various issues [7, 11, 12]. One of them is the utilization of an improper uniform color space. Another issue is the assumption that the used reference sources render the colors perfectly is not always correct e.g., at very low and very high CCTs. These problems cause inaccurate results especially for the light sources having saturated color components. In addition to this, the arithmetic mean used during the calculation of CRI allows for the compensation of a low CRI value belonging to a certain test sample by the high CRIs of other test samples.

These problems of CRI are later addressed by Davis and Ohno who introduced the color quality scale (CQS) as an alternative to CRI [7]. CQS and CRI both employ the same reference sources. However, the CQS makes use of fifteen commercially available Munsell samples, all having highly saturated colors. This selection is based on the observation that a light source successfully rendering the saturated colors also successfully renders the unsaturated colors successfully [7]. This is especially important for the narrow-band emitters such as LED and nanocrystal-based light sources. Different than CRI, CQS employs the L*a*b* color space, which is a more uniform color space compared to (u, v) color space. Another improvement in CQS compared to CRI is the addition of a saturation factor that neutralizes the effect of increasing the object chroma under the test illuminant with respect to a reference source. Furthermore, CQS does not allow the compensation of a poorly rendered test source by other successfully rendered sources by calculating the root-mean-square of individual color differences. Another fine-tuning in CQS compared to CRI is the change of the scale from the range of -100 to 100 to the range of 0 to 100. Finally, in CQS a correction for the low CCTs is introduced, and the final value of the CQS is determined.

The calculation of CQS employs 15 Munsell test samples whose reflection spectra we provide in Fig. 4.3 and tabulate in Appendix A of this brief.

An important difference of CQS compared to CRI is the reference light source, which is assumed to render the real colors of the objects perfectly. If the correlated color temperature of the test source is less than 5000 K, the reference source is the usual blackbody radiator. In the case that the correlated color temperature is between 5000 and 7000 K, the reference light source is calculated using Eqs. (4.14)–(4.18) as follows:

$$x = -4.7070 \times 10^9/T^3 + 2.9678 \times 10^6/T^2 + 0.09911 \times 10^3/T + 0.244063 \tag{4.14}$$

$$y = 3x^2 + 2.87x - 0.275 \tag{4.15}$$

$$m_1 = \frac{-1.3515 - 1.7703x + 5.9114y}{0.0241 + 0.2562x - 0.7341y} \tag{4.16}$$

$$m_2 = \frac{0.03 - 31.4424x + 30.0717y}{0.0241 + 0.2562x - 0.7341y} \tag{4.17}$$

$$R(\lambda) = D_1(\lambda) + m_1 D_2(\lambda) + m_2 D_3(\lambda) \tag{4.18}$$

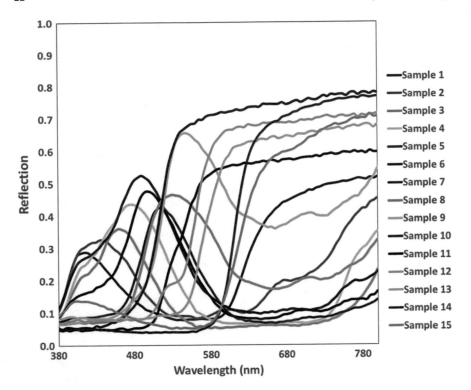

Fig. 4.3 Reflection spectra of 15 Munsell samples used in the calculation of CQS

where T stands for the correlated color temperature, D_i stands for the ith CIE standard daylight illuminants whose spectral power distributions are provided in Appendix A of the brief.

In the case that the correlated color temperature of the test light source is more than 7000 K, x is modified using Eq. (4.19):

$$x = -2.0064 \times 10^9/T^3 + 1.9018 \times 10^6/T^2 + 0.24748 \times 10^3/T + 0.23704 \tag{4.19}$$

Next, the intensities of the reference and test source are scaled such that their Y chromaticity coordinates become 100.

After calculating the reference source and scaling both reference and test sources, we are now ready to calculate the differences of the reflected colors when Munsell samples are illuminated with the reference and test sources. For this purpose, the reflected spectra $q_{ref,i}$ and $q_{test,i}$ from a Munsell sample i illuminated by the reference and test sources, respectively, are calculated as follows:

$$q_{ref,i}(\lambda) = r_i(\lambda)R(\lambda) \tag{4.20}$$

$$q_{test,i}(\lambda) = r_i(\lambda)s(\lambda) \tag{4.21}$$

where $R(\lambda)$ and $s(\lambda)$ are the reference and test sources, respectively, whose Y values were scaled to 100. These reflection spectra are then used to calculate the L*a*b* coordinates for both $q_{ref,i}$ and $q_{test,i}$ where the nominal white source is selected as the reference source $R(\lambda)$. An important point here is that L*a*b* coordinates of the $q_{test,i}$ are calculated after carrying out chromatic adaptation transformation to the test illuminant using CMCCAT2000 method. The inputs of this transformation are the X, Y and Z tristimulus values of (1) $q_{test,i}(\lambda)$ (whose Y is set to 100), (2) test source $s(\lambda)$, (3) adapting white source $R(\lambda)$ (whose Y is set to 100), (3) adapting background luminance set to 1000, and (4) surround luminance set to 1000. Based on these calculated L*a*b* coordinates, the saturation difference of the reflected color $\Delta C_{ab,i}$ from sample i between the $q_{ref,i}(\lambda)$ and $q_{test,i}(\lambda)$ are found using Eq. (4.22):

$$\Delta C_{ab,i} = \sqrt{a_{ref,i}^2 + b_{ref,i}^2} - \sqrt{a_{test,i}^2 + b_{test,i}^2} \tag{4.22}$$

Subsequently, the L*a*b* Euclidian color difference ΔE_i between $q_{test,i}(\lambda)$ and $q_{ref,i}(\lambda)$ is found as shown below:

$$\Delta E_i = \sqrt{\left(L_{ref,i} - L_{test,i}\right)^2 + \left(a_{ref,i} - a_{test,i}\right)^2 + \left(b_{ref,i} - b_{test,i}\right)^2} \tag{4.23}$$

The corrected color difference $\Delta E_{c,i}$ becomes $\Delta E_{c,i} = \Delta E_i - \Delta C_{ab,i}$ if $\Delta C_{ab,i}$ is greater than zero, otherwise $\Delta E_{c,i}$ becomes equal to ΔE_i. The total color difference is found by finding the root mean square of the corrected color differences as expressed in Eq. (4.24):

$$\Delta E_{rms} = \sqrt{\frac{1}{15} \sum_{i=1}^{15} \Delta E_{c,i}^2} \tag{4.24}$$

An important improvement of CQS over CRI is the introduction of a correlated color temperature factor. Finding this factor requires the calculation of the gamut area F_{total} for each Munsell sample i (if $i = 15$, $i + 1$ is assumed to be 1). The calculation is carried out using Eqs. (4.25)–(4.30):

$$A_i = \sqrt{a_i^2 + b_i^2} \tag{4.25}$$

$$B_i = \sqrt{a_{i+1}^2 + b_{i+1}^2} \tag{4.27}$$

$$C_i = \sqrt{(a_{i+1} - a_i)^2 + (b_{i+1} - b_i)^2} \tag{4.27}$$

$$t_i = \frac{A_i + B_i + C_i}{2} \tag{4.28}$$

$$F_i = \sqrt{t_i(t_i - A_i)(t_i - B_i)(t_i - C_i)} \tag{4.29}$$

$$F_{total} = \sum_{i=1}^{15} F_i \tag{4.30}$$

If F_{total} is greater than 8210 K, the correlated color temperature factor f_{CCT} becomes 1, otherwise f_{CCT} is $F_{total}/8210$. Finally, the CQS is calculated using Eq. (4.31):

$$CQS = 10 \log\left(e^{\frac{100-3.105 \times \Delta E_{rms}}{10} + 1}\right) \times f_{CCT} \tag{4.31}$$

4.3   Photometry: Stimulus Useful for the Human Eye

The first pair of radiometric-photometric quantities that we introduce here is the radiant and luminous flux. Radiant flux is basically the power radiated by a light source and has units of W_{opt}. The luminous flux (Φ), on the other hand, is defined as the useful optical radiation for the human eye, expressed in units of lumen (lm), and calculated by using Eq. (4.32) where $P_R(\lambda)$ and $V(\lambda)$ stand for the spectral radiant flux and the photopic eye sensitivity function, respectively.

$$\Phi = 683 \frac{lm}{W_{opt}} \int P_R(\lambda)V(\lambda)d\lambda \tag{4.32}$$

Another important radiometric quantity is the irradiance, which is the optical power per unit area and expressed in units of W_{opt}/m^2. The illuminance is the irradiance subject to the photopic human eye sensitivity function, and it has units of lm/m^2 or equivalently lux. Given the spectral irradiance $P_I(\lambda)$, the illuminance (IL) is expressed as in Eq. (4.33). The illuminance is a quantity which is used to assess the effect of the lighting on the human circadian cycle.

$$IL = 683 \frac{lm}{W_{opt}} \int P_I(\lambda)V(\lambda)d\lambda \tag{4.33}$$

Among the most important pairs of radiometric-photometric quantities we can include are the radiance and luminance. The radiance that is expressed in $W_{opt}/(m^2 sr)$ is the optical power per solid angle per unit area. For a spectral radiance $P_L(\lambda)$, the luminance L that is the optical radiance useful to human eye is found in units of $lm/(m^2 sr)$, or equivalently cd/m^2 using Eq. (4.34). The calculation makes use of photopic eye sensitivity function as given by

$$L = 683 \frac{lm}{W_{opt}} \int P_L(\lambda)V(\lambda)d\lambda \tag{4.34}$$

where $V(\lambda)$ is the photopic eye sensitivity function.

Although the luminance levels are traditionally calculated using photopic eye sensitivity function, there is a need to quantitatively express accurate luminance levels in different visual regimes, especially for the mesopic vision regime, which corresponds to the road lighting conditions. In 2010, CIE addressed this problem by publishing a recommended system called CIE 191:2010. According to this recommendation, the mesopic vision regime falls into any photopic luminance levels between 0.005 and 5 cd/m^2. When the luminance level is below 0.005 cd/m^2, the vision regime is considered to be the scotopic regime while the luminance greater than 5 cd/m^2 corresponds to the photopic vision regime [13]. The mesopic luminance L_{mes} is found using Eq. (4.35) where $V_{mes}(\lambda)$ is the mesopic eye sensitivity function whose maximum value is 1, λ_0 is 555 nm, and $P(\lambda)$ is the spectral radiance.

$$L_{mes} = 683/V_{mes}(\lambda_0) \int P(\lambda)V_{mes}(\lambda)d\lambda \tag{4.35}$$

The mesopic eye sensitivity function is suggested to be a linear combination of the photopic and scotopic eye sensitivity functions, calculated using Eq. (4.36) where $V(\lambda)$ and $V'(\lambda)$ stand for the photopic and scotopic eye sensitivity functions, respectively. M(m) is a normalization constant equating the maximum value of $V_{mes}(\lambda)$ to 1, and m is the coefficient that sets the contribution of scotopic and photopic eye sensitivity functions according to visual adaptation conditions.

$$M(m)V_{mes}(\lambda) = mV(\lambda) + (1-m)V'(\lambda) \tag{4.36}$$

Here m is 0 if L_{mes} is greater 5 cd/m^2, and m is 1 if L_{mes} is smaller than 0.005 cd/m^2. The intermediate values of m and L_{mes} are found using an iterative approach employing the relations in Eqs. (4.37) and (4.38) and setting m_0 to 0.5.

$$L_{mes,n} = \frac{m_{n-1}L_p + (1-m_{n-1})L_s V'(\lambda_0)}{m_{n-1} + (1-m_{n-1})L_s V(\lambda_0)} \tag{4.37}$$

$$m_n = a + b\log_{10}\left(L_{mes,n}\right) \tag{4.38}$$

where a and b are 0.7670 and 0.3334, respectively, n is the step of iteration, m_n is always between 0 and 1, L_s and L_p are the scotopic and photopic luminances, respectively, and $V(\lambda_0)$ and $V'(\lambda_0)$ are the photopic and scotopic eye sensitivity function values at 550 nm. The iteration is continued until the difference between m_n and m_{n-1} becomes negligibly low.

From the device point of view, achieving the desired luminance levels is important. However, this is just one part of the performance, also the efficiency of the light-emitting devices should be high. There are two metrics that need to be considered while designing an efficient light source. The first one is the optical efficiency of the device. It basically evaluates how efficiently the radiated light can be perceived by the human eye. This metric is called the luminous efficacy of the optical radiation (LER), which is calculated using Eq. (4.39). In this equation, P(λ) stands for the spectral radiation and V(λ) is the eye sensitivity function at the vision regime of interest.

LER has units of lm/W_{opt} and takes a maximum value of 683 lm/W_{opt}, which can only be achieved by a monochromatic light source emitting at 555 nm. An excellent white light source should have LER > 350 lm/W_{opt} [10].

$$LER = \frac{683 \, lm/\, W_{opt} \int P(\lambda)V(\lambda)d\lambda}{\int P(\lambda)V(\lambda)d\lambda} \tag{4.39}$$

The second efficiency metric evaluates how efficiently the sources radiate light per supplied electrical power. This metric that disregards the human perception specifications is called the wall plug efficiency or power conversion efficiency, which is essentially the total collected optical power divided by electrical power. When we consider the human perception, on the other hand, the efficiency metric should include the luminous flux. The resulting quantity is known as the luminous efficiency (LE), computed using Eq. (4.40) where $P(\lambda)$ is the spectra radiance and P_{elect} is the electrical power. The unit of LE is lm/W_{elect}. Today, the LEs of the efficient light sources are in the proximity of 150 lm/W_{elect} [14].

$$LE = \frac{683 \, lm/\, W_{opt} \int P(\lambda)V(\lambda)d\lambda}{P_{elect}} \tag{4.40}$$

References

1. Schubert EF (2009) Light-emitting diodes, 2nd edn. Cambridge University Press
2. Raynham PST (2003) White light and facial recognition. Light J 68:29–33
3. Thornton WA (1972) Color-discrimination index. J Opt Soc Am
4. Xu H (1993) Color-rendering capacity of light. Color Res Appl
5. Hashimoto K, Nayatani Y (1997) Visual clarity and feeling of contrast. Color Res Appl
6. Judd DB (1967) A flattery index for artificial illuminants. Illum Eng. J: 593–598
7. Davis W, Ohno Y (2010) Color quality scale. Opt Eng 49(3):033602
8. "CIE publication No. 15—Colorimetry," 1971
9. "CIE publication No. 13.3—Method of measuring and specifying color-rendering of light sources," 1995
10. Erdem T, Demir HV (2013) Color science of nanocrystal quantum dots for lighting and displays. Nanophotonics 2(1):57–81
11. Ohno Y, Davis W (2010) Rationale of color quality scale. Energy: 1–9
12. Ohno Y (2013) Color quality of white LEDs. Top Appl Phys 126:349–371
13. Eloholma M et al (2005) Mesopic models—from brightness matching to visual performance in night-time driving: a review. Light Res Technol 37(2):155–175
14. Erdem T, Demir HV (2016) Colloidal nanocrystals for quality lighting and displays: milestones and recent developments. Nanophotonics 5(1):74–95

Chapter 5
Common White Light Sources

Abstract In this Chapter we describe the features of common light sources. We first present the spectral features of the sun and discuss its colorimetric properties. Next, we summarize the properties of traditional light sources including incandescent lamps, fluorescent lamps, and high-pressure sodium lamps. Subsequently, we discuss the white light-emitting diodes of various types.

Keywords Sun · Incandescent lamps · Fluorescent lamps · High-pressure sodium lamps · LEDs

We would like to start this section by summarizing the desired features of an artificial white light source. A good white light source should be able to render the real colors of the objects as good as possible. In other words, they should have high color rendering index or color quality scale values. Also the radiated intensity should overlap with human eye sensitivity function, which can be assessed using luminous efficacy of optical radiation. LER values close to 350 lm/W_{opt} can be considered photometrically efficient. It is here worth pointing out that the eye sensitivity function depends on the vision regime, which may vary according to the desired medium of working for the light source. In addition to these, a warmer white shade is desirable especially for indoor lighting to avoid disturbing the human biological rhythm. Quantitatively, this corresponds to correlated color temperatures (CCTs) smaller than 4000 K. Another critical parameter is of course the electrical efficiency of the light sources, which can be evaluated using luminous efficiency. Nowadays, the technology enables LE values above 100 lm/W_{elect}, nevertheless, methods and processes are being developed to go beyond 150 lm/W_{elect}.

5.1 The Sun

Before proceeding to the properties of artificial light sources, here we first would like to start with the sun, which has been the ultimate light source for all the terrestrial livings. Its spectral features have affected the evolutionary dynamics of our body. The average irradiance reaching the surface of earth is around 1 kW/m^2 [1]. At these

© The Author(s), under exclusive licence to Springer Nature Singapore Pte Ltd. 2019   27
T. Erdem and H. V. Demir, *Color Science and Photometry for Lighting with LEDs and Semiconductor Nanocrystals*, Nanoscience and Nanotechnology,
https://doi.org/10.1007/978-981-13-5886-9_5

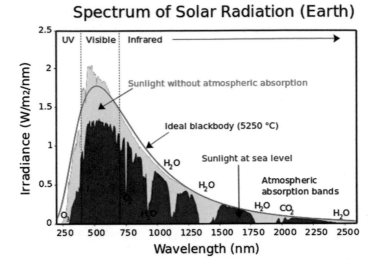

Fig. 5.1 Spectral irradiance of sun at the sea level and outside the atmosphere. Ref. [2]

energy levels, the human eye is obviously in the photopic regime where the cones dominate the vision. In the photopic regime, the average sensitivity of the eye peaks at 550 nm, which is also the value where the atmosphere's transparency is the highest. Also a blackbody radiator at a temperature of 5250 K peaks at this wavelength which is the horizon color temperature of the sun light.

Since the sun can be classified as a thermal radiator, its radiation spectrum extends to long wavelengths where the human eye is no longer sensitive (see Fig. 2.3), which makes replicating the sun spectrum in its entirety while designing a light source a bad idea in terms of the efficiency. Nevertheless, generating a broad-band light spectrum without thermal emission has been tricky and thus, utilizing blackbody radiators have been preferred for years (Fig 5.1).

5.2 Traditional Light Sources

Among traditional light sources, the incandescent lamps are the ones whose radiation resembles the sun light most as blackbody radiators. The light emission mechanism in these lamps relies on heating a wire—typically made of tungsten—by applying electric current. The temperature of the wire reaches couple of thousands of Kelvins, possibly causing oxidation and evaporation of metal filaments. To minimize these effects, modern incandescent lamps are filled with inert gas. Being a thermal radiator, its emission spectra follow the characteristics of a standard blackbody radiator (Fig. 5.2). Because blackbody radiators are considered as perfect color renderers, the color rendering index (or color quality scale) of incandescent lamps is 100. On

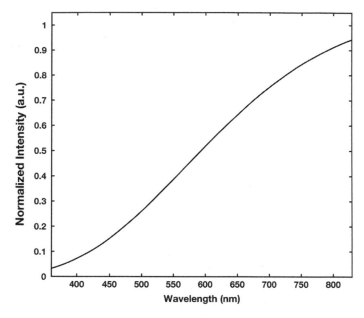

Fig. 5.2   The emission spectrum of an incandescent lamp at a CCT of 3000 K

the other hand, the overlap of their spectra with the eye sensitivity function remains low due to the long tail reaching far infrared. This unavoidably decreases the luminous efficacy values to ca. 8–24 lm/W_{opt} [3]. Since the thermal radiation process is an extremely inefficient way of producing light, the power conversion efficiency of these lamps remains only at ca. 5%, which significantly drops the luminous efficiency. This is the main reason behind the current ban of these lamps in various parts of the world [4–6].

Before the era of LEDs, the fluorescent lamps were the most efficient white light sources in the market. Even today, a significant portion of the energy-saving lamps on the market are fluorescent lamps. These lamps basically rely on the color conversion of ultraviolet light produced by low pressure mercury vapor through electric discharge mechanism. The short-wavelength radiation of mercury vapor is absorbed by the phosphors and converted to broad-band white light (see Fig. 5.3 for a typical spectrum). Modern fluorescent lamps use phosphors made of rare-earth ions as color converters. These phosphors have a high photoluminescence quantum efficiency (i.e., number of photons emitted per absorbed excitation photon) helping them to reach luminous efficiency values up to 100 lm/W_{elect}. Depending on the composition of the phosphors, they can also have high color rendering index values and a warm white shade. Nevertheless, the lack of color tunability using phosphors does not enable to optimize all these performance parameters at the same time. Other important concerns regarding these lamps are the mercury content making their waste hazardous and the ultraviolet radiation stemming from the discharge of mercury vapor. Because

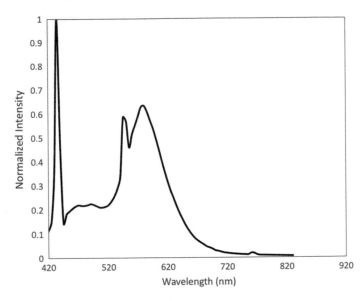

Fig. 5.3 A typical cool white fluorescent lamp emission spectrum

of these shortcomings, alternatives for these lamps are in demand and today light-emitting diodes step forward as an important replacement.

Another widely used white light source is the high-pressure sodium lamps. Similar to fluorescent lamps, these lamps are also gas discharge lamps. At low pressure, sodium vapor emits monochromatic light close to 590 nm, while broadband emission is observed as the pressure increases (Fig. 5.4). The white light of high pressure lamps comes from sodium-mercury amalgam. The yellow and turquoise light stems from the high-pressure sodium (HPS) whereas the rest of the emission comes from the mercury vapor. An important characteristic of sodium lamps is their high luminous efficiencies reaching 150 lm/W$_{elect}$. Nevertheless, they have very low color rendering index making them less desirable for indoor lighting applications. Furthermore, these lamps cannot provide high luminance in mesopic vision conditions. This is due to their very low correlated color temperatures <3000 K because of strong yellow content. These lamps are often used in road lighting, probably because of their energy efficiency and high luminous flux measured in photopic regime. However, the visual regime of road lighting is different than photopic regime and thus, HPS lamps are far from satisfying quality lighting conditions.

5.3   White LEDs

There are two main methods of obtaining white light emission using light-emitting diodes (LEDs). The first approach is the multi-chip LED approach, where the white

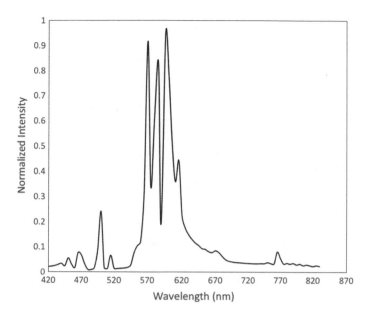

Fig. 5.4 Emission spectrum of a typical HPS lamp

light emission is generated via the collective radiation of several LED chips (at least three of them for red, green, and blue). The second method makes use of the photoluminescence of a fluorescent material integrated on top of an LED chip emitting at a shorter wavelength. This phenomenon is known as the color conversion, wavelength up-conversion, or energy down-conversion.

5.3.1 Multi-chip Approach

In multi-chip approach, LEDs emitting in different colors are combined to obtain white light. Modern green and blue LEDs are fabricated using GaN/InGaN material system while AlGaInP is employed to realize high-quality red LEDs [7]. In Fig. 5.5, we provide typical emission spectra of LEDs of different colors. These emission bandwidths of LEDs are typically around 20–25 nm. Owing to this narrow linewidth, all the colorimetric and photometric features can be optimized simultaneously by carefully combining LEDs of various colors. However, their main drawback is the cost of the electrical circuitry to drive these LEDs. Another important problem is the lack of material system enabling efficient green emission, which is referred to as the green gap problem. Although the current green and yellow LEDs still remain behind their blue and red counterparts, novel materials and techniques are widely being investigated to overcome this green gap.

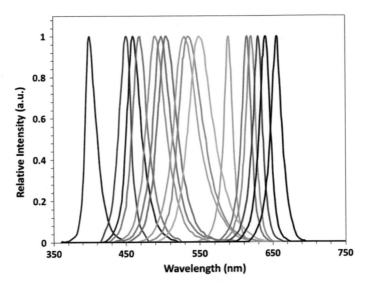

Fig. 5.5 Spectra of various LED chips with peak emission wavelengths ranging from 400 to 655 nm

5.3.2   Color Conversion Approach

Different than the multi-chip approach, the color conversion method relies on the excitation of color-converting materials by higher energy photons provided by a short-wavelength LED. Typically blue LED chips emitting around 450–460 nm and alternatively near-ultraviolet LED chips emitting around 380–400 nm are employed as the high-energy photon source. As color converters, most commonly rare earth ion doped phosphors, and more recently colloidal quantum dots are used.

Alternatively, there are several organic dyes and polymers that can be incorporated in LEDs [7, 8]. Despite their high quantum efficiencies, their narrow-band absorption limits the choice of the pump LED, and their low photostability is an important bottleneck for their use in commercial devices. Moreover, these organic materials usually have broad-band emission; thus, simultaneous optimization of all the colorimetric and photometric criteria becomes challenging.

5.3.3   Broadband Versus Digital Color Lighting

The rare-earth ion based phosphors are the most ubiquitously used color converters in modern white LEDs. They are incorporated into a garnet such as yttrium aluminum garnet (YAG) doped with rare-earth ions such as gadolinium (Gd), cerium (Ce) and terbium (Te) ions [9]. The emission spectrum of a typical white LED using phosphor color converters is presented in Fig. 5.6. As apparent in this figure, phosphors

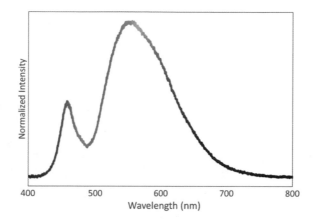

Fig. 5.6 Emission spectrum of a color-converting LED using phosphors made of rare-earth ions

typically possess broad emission spectra enabling the creation of white light when hybridized with a blue LED. This broadness, however, causes a photometric problem as the tail of the emission spectrum reaches the near infrared region where the human eye is not sensitive anymore. This inevitably decreases the luminous efficacy of optical radiation. To overcome this problem, the spectra of these phosphors are tuned by controlling the ion concentration and composition so that the overlap with the human eye sensitivity function is maximized as much as possible. Combined with their near unity quantum efficiencies, this results in high luminous efficiency values reaching >100 lm/W$_{\text{elect}}$. However, the ability to tune the emission spectra of these phosphors is very limited; therefore, optimizing all the colorimetric and photometric parameters at the same time is not possible.

All the color converters explained above share a common problem: The difficulty in tuning the emission spectra in order to optimize colorimetric and photometric performance. Reaching this goal is only possible when strategically designed combinations of narrow-band emitters are used. Among the candidates of narrow-band emitters, quantum dots have an important place [10, 11]. The chemical synthesis techniques enable precise control over their size and the particle size distribution governs the emission color and linewidth of their emission. Currently, all the visible regime can be spanned with these materials while the bandwidths down to 20 nm can be realized at the same time (see Fig. 5.7). Furthermore, the photoluminescence quantum efficiencies close to unity have been obtained. Nevertheless, these quantum dots are usually made of Cd-containing components such as CdSe and CdS, which raises questions regarding their widespread use. Although Cd-free quantum dots have been successfully synthesized, their quantum efficiencies remain low and emission linewidths are broader compared to their Cd-containing counterparts. In recent years a special class of these materials have attracted great interest in optics and photonics. These are known as perovskite nanocrystals because of their crystal structures. These materials also possess high quantum efficiencies and narrow emission spectra; however, their low stability is a major challenge in addition to the Pb content in efficient perovskite emitters. As an alternative, colloidal quantum wells

Fig. 5.7 Normalized
photoluminescence (PL)
spectra of semiconductor
nanocrystal quantum dots of
varying emission colors

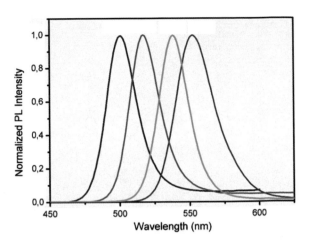

of inorganic semiconductors have emerged within the last couple of years. Their most important feature is the extraordinarily narrow emission reaching linewidths as narrow as 8 nm. This narrow emission should enable fine tuning of the spectra; however, the biggest challenges avoiding their widespread use are (i) the inability to obtain their efficient solid films without significantly altering the optical properties, (ii) difficulties in realizing efficient quantum wells at any desired wavelength in the visible regime, and (iii) the heavy metal content (Cd) of efficient colloidal quantum wells.

References

1. "Tutorial: introduction to solar radiation." [Online]. Available: https://www.newport.com/t/introduction-to-solar-radiation. Accessed 30 Jul 2018
2. "Sunlight." [Online]. Available: https://en.wikipedia.org/wiki/Sunlight. Accessed 12 Nov 2018
3. Agrawal DC, Leff HS, Menon VJ (1996) Efficiency and efficacy of incandescent lamps. Am J Phys 64(5):649–654
4. Howarth NAA, Rosenow J (2014) Banning the bulb: institutional evolution and the phased ban of incandescent lighting in Germany. Energy Policy 67:737–746
5. Mills B, Schleich J (2014) Household transitions to energy efficient lighting. Energy Econ 46:151–160
6. Schielke T (2007) Energy efficiency with lamp technology. Energy efficiency and visual comfort through intelligent design. ERCO Lichtbericht 83
7. Schubert EF (2006) Light-emitting diodes, 2nd edn. Cambridge University Press
8. Guha S, Bojarczuk NA (1998) Multicolored light emitters on silicon substrates. Appl Phys Lett 73(11):1487–1489
9. Graydon O (2011) The new oil? Nat Photonics 5(1):1
10. Erdem T, Demir HV (2013) Color science of nanocrystal quantum dots for lighting and displays. Nanophotonics 2(1):57–81
11. Erdem T, Demir HV (2016) Colloidal nanocrystals for quality lighting and displays: milestones and recent developments. Nanophotonics 5(1):74–95

Chapter 6
How to Design Quality Light Sources With Discrete Color Components

Abstract White light sources using discrete emitters require careful design and optimization. The first step of the design should be determining the intended use of the light source so that application specific requirements can be addressed. Subsequently, optimal designs made of discrete emitters should be determined and finally, experimental implementation of the light source should be carried out. In this Chapter of the brief, we limit ourselves to the use of discrete emitters for indoor and outdoor lighting together with display backlighting applications. For each application, we summarize the requirements that need to be satisfied and present design guidelines to implement quality light sources made of discrete emitters.

Keywords Optimization of photometric quantities · Indoor lighting · Outdoor lighting · Display backlighting

6.1 Advanced Design Requirements for Indoor Lighting

For indoor lighting applications it is essential for a light source to reflect the real colors of objects. This requires CRI and/or CQS to be maximized. Furthermore, the light source needs to have a warm white shade for a comfortable vision, necessitating CCTs lower than 4500 K (preferably < 4000 K). Also, the overlap with the human eye sensitivity function should be maximized, which is quantified by LER. To ensure energy saving, LE of the device should be maximized as well. Ideally, for LER we expect to reach values greater than 350 lm/W_{opt} while for the LE targeted levels should be >100–120 lm/W_{elect} to compete with the existing light sources. In this part of the brief, we summarize the design strategies in the light of Refs. [1] and [2] to optimize all the parameters for a high-quality light-emitting device. We start with the necessary conditions to realize high-quality white light spectrum for indoor lighting and then continue with summarizing the requirements for high device efficiency.

Optimizing the spectral features of white light sources employing discrete emitters is a complicated task. To have a qualitative picture while approaching this problem, one needs to have an idea on the trade-offs between several performance metrics of interest. In Ref. [1], this problem was addressed by calculating the performance of

© The Author(s), under exclusive licence to Springer Nature Singapore Pte Ltd. 2019
T. Erdem and H. V. Demir, *Color Science and Photometry for Lighting with LEDs and Semiconductor Nanocrystals*, Nanoscience and Nanotechnology,
https://doi.org/10.1007/978-981-13-5886-9_6

Fig. 6.1 CRI versus LER
trade-off for
nanocrystal-integrated white
LEDs at different CCTs.
Reproduced with permission
from Ref. [4] © Science
Wise Publishing &
DeGruyter 2013

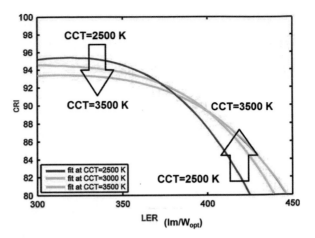

modelled white LED spectra made of colloidal nanocrystal quantum dots using real-istic properties. The first point that needs to be clarified is the number of color com-ponents. It turns out that when narrow emitters such as quantum dots are employed, utilizing only three color components (i.e., blue, green, and red) cannot provide suffi-cient degrees of freedom to optimize photometric properties and achieve high-quality lighting. As explained in Ref. [1] and also stated by Tsao [3], the minimum number of color components needed is four which can be blue, green, yellow, and red. For a good white light source, a good white light source should be able to render the real colors of the objects. When narrow emitters such as colloidal quantum dots or individual LED chips that have emission linewidths between 20–50 nm are employed and when only three colors or less are used, the spectrum of the light source cannot sufficiently span the whole visible regime leading to poor color rendition perfor-mance which translates to low CRI or CQS values. Nevertheless, it turns out that employing four color components is enough to solve this problem and achieving good color rendition is possible while also optimizing other parameters.

The analyses presented in Ref. [1] also show that there is a fundamental trade-off between LER and CRI. At a fixed CCT value, as a general trend, the increase of LER is accompanied by a decrease in the CRI and vice versa (Fig. 6.1). An interesting finding is that the maximum CRI values that can be obtained at a given LER favour warmer white lights until ~370 lm/W_{opt} whereas after this value high CRI values can be obtained at the expense of cooler white shades. All in all, below ~370 lm/W_{opt}, CRIs > 92 are feasible. However, beyond ~370 lm/W_{opt}, it is not possible to sustain CRI levels above 90.

The emission wavelength, relative amplitude, and linewidth of each color compo-nent has a profound effect on the performance of the designed white light source. The results obtained from simulations are presented in Fig. 6.2. As seen in this figure, high-quality lighting depends strongly on the properties of the red color compo-nent. The peak emission wavelength of this color component should be at 620 nm in photopic regime. The designer does not have a large flexibility to play with this

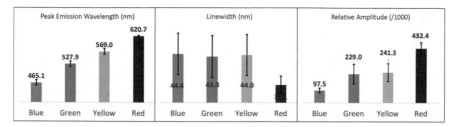

Fig. 6.2 Average and standard deviations of the peak emission wavelength, linewidth and relative amplitude of nanocrystal quantum dot color components to obtain white LED spectra possessing CRI > 90, LER > 380 lm/W$_{opt}$, and CCT < 4000 K

value due to the low standard deviation shown in Fig. 6.2. Furthermore, this color component should be the most dominant one in the white light spectrum. Second important condition on high-quality lighting is imposed by the blue component. Due to warm white light requirement, the amplitude of the blue color component has to be strictly placed around 90/1000. As in the case of red color component, the designer does not again have the flexibility to change this value without sacrificing from the overall performance. On the other hand, the computations indicate there is no major restriction on the linewidth of this color component. Furthermore, the peak emission wavelength can be flexibly chosen around 465 nm as the large standard deviations in Fig. 6.2 imply. It turns out that the designer has the highest flexibility when choosing the green and yellow color components compared to the blue and the red. The simulations show that the peak emission wavelengths of these colors can be selected around 528 and 569 nm without a strong restriction as the corresponding large standard deviations indicate. Furthermore, both broad and narrow emission spectra of green and yellow components are found to allow for high performance. Finally, the relative amplitudes of 229/1000 and 241/1000 with standard deviations > 70/1000 for green and yellow components, respectively, mean that the amplitudes of these colors should be between the amplitudes of the blue and red color components. It turns out that the designer has a large flexibility to play with these parameters without sacrificing the end performance of the device. The white LED spectrum generated using these average values possesses a CRI of 91.3, an LER of 386 lm/W$_{opt}$, and a CCT of 3041 K (Fig. 6.3). This shows that extremely high-quality artificial white light can be generated via color conversion by employing nanocrystal quantum dots integrated with LEDs as optical pumps when the requirements listed above are addressed.

After establishing the requirements for high-quality white light spectrum generation, the next essential question of LED design is how to place the nanocrystal quantum dots on top of the blue LED, and what the performance limits in terms of energy consumption are. In Ref. [2], the answers to these questions are sought through a computational approach. In this work, the potential luminous efficiency of LEDs deploying two types of nanocrystal quantum dot color converting thin film architectures are investigated. These are (i) three separate layers of nanocrystal quan-

Fig. 6.3 Designed spectrum of nanocrystal quantum dot integrated white LED generated using the results in Fig. 6.2 along with a summary of its performance in the inset. Reproduced with permission from Ref. [4]

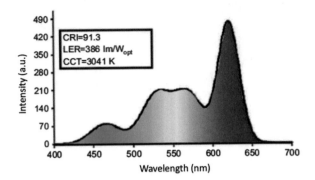

tum dots (with the green first, followed by the yellow, and then the red on top) and (ii) the nanocrystal blended together to form a single coating layer on a blue LED.

To investigate the ultimate limits, the hypothetical case of quantum dot films having unity quantum efficiency is simulated. It turns out that the energy loss due to photon energy down-conversion is at least 17% when a photometrically efficient design is employed. This corresponds to a luminous efficiency of 315 lm/W_{elect} when a perfect blue LED chip with a power conversion efficiency of 100% is assumed. This means a highly efficient blue LED having a power conversion efficiency of 81.3% is used [5], the maximum obtainable luminous efficiency becomes 256 lm/W_{elect}. When a typical blue LED efficiency of 40% is considered, then the maximum obtainable LE becomes 126 lm/W_{elect}. This shows that one of the most critical conditions to achieve a highly efficient design is to realize highly efficient blue LED chips.

Another important question is how to construct the nanocrystal thin films to realize high efficiency. The results of Ref. [2] show that the layered architecture outperforms the blended architecture in terms of overall luminous efficiency. In order to achieve high photometric performance together with a luminous efficiency level of 100, 150, and 200 lm/W_{elect}, nanocrystals quantum dots need to have quantum efficiency of at least 43, 61, and 80%, respectively, if an LED with a power conversion efficiency of 81.3% is employed. On the other hand, when utilizing the blended quantum dot white LED architecture, the required quantum efficiencies of the QDs need to increase to 47, 65, and 82%, respectively. This is essentially what one should expect. In a layered architecture where the green nanocrystals are on top and the red nanocrystals are at the bottom, the energy loss due to color conversion is minimized compared to the blend architecture. If the order of the layered nanocrystals is reversed, the resulting efficiency remains below that of the blend film due to increased radiative energy transfer from nanocrystals emitting high energy photons to nanocrystals emitting low energy photons.

6.2 Advanced Design Requirements for Outdoor Lighting

Different than indoor lighting, for outdoor lighting the important point is to achieve a lighting condition that can help the pedestrians and drivers recognize their surroundings. For that purpose, the perceived luminance plays a critical role while the shade of the white and color rendition performance are not as critical as in the case of indoor lighting. Nevertheless, in the literature a light source having a color rendering index >80 was found to increase the visual acuity under street lighting conditions. The critical difference to consider here is the change in the eye sensitivity function under the mesopic visual regime conditions where the rods and cones contribute to the vision together.

To take this variation into account, in Ref. [6] the design requirements for achieving high-quality lighting with nanocrystal quantum dots are established. This study starts with determining the luminance levels to satisfy the existing road lighting standards, which vary significantly depending on the country. For this purpose, four different luminance levels were selected such that the British [7] and US [8] standards can be fulfilled. These are 0.50, 0.80, 1.25, and 1.75 cd/m^2 which are named in Ref. [6] as Mesopic 1, 2, 3, and 4, respectively. Next, the performance of the commercial white light sources was evaluated. Among a cool white fluorescent lamp (CWFL), an incandescent lamp with a CCT of 3000 K, a metal-halide lamp (MH), a high-pressure sodium lamp (HPS), and a mercury vapor lamp (MV), the CWFL was found the most efficient commercial light source for Mesopic 1 and 2 standards, whereas the HPS becomes the most efficient source for Mesopic 3 and 4 standards. To test the potential of nanocrystal quantum dot integrated white LEDs, spectra having the same radiance as the CWFL has for Mesopic 1 and 2 standards and HPS has for Mesopic 3 and 4 are generated. Among ca. 100 million spectra tested, the ones possessing CRI or CQS > 85 and higher luminance than the commercial light source are investigated. Results indicate that using nanocrystal quantum dots, it is possible to simultaneously obtain high mesopic luminance as well as successful color rendition performance and warm white appearance. Having a blue component close to 460 nm and a red component around 610 nm turn out to be crucial for realizing this performance (Fig. 6.4). In addition, the weight of the blue component should be around 150/1000, whereas the relative intensity of the red component should be chosen around 450/1000. Different than the indoor lighting conditions, the outdoor lighting requirements do not pose strong limitations on the emission linewidths of each color component. Results summarized in Fig. 6.4 show that bandwidths of 35–40 nm can lead to the generation of high-quality road lighting.

Maintaining high electrical efficiency is another important concern while designing a white light source. To answer what efficiency levels can make the nanocrystal quantum dot integrated LEDs more efficient than the commercial light sources, a systematic calculation has been carried out in Ref. [6]. Assuming that the power conversion efficiencies of CWFL as 28%, HPS as 31%, and MH as 24%, the quantum dot integrated LEDs satisfying high quality road lighting conditions can make

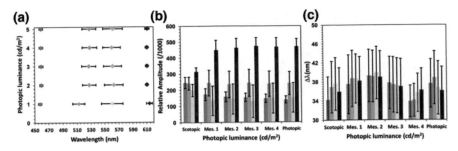

Fig. 6.4 Average and standard deviation (bars) of (**a**) peak emission wavelength, relative amplitude, and (**c**) full-width at half maximum belonging to the spectra suitable for road lighting conditions. Colors in the graph indicate the corresponding color components. Reproduced with permission from Ref. [6] © Science Wise Publishing & DeGruyter 2014

more efficient street lighting sources if the final power conversion efficiency of the device is larger than 24%, which was also shown to be experimentally feasible [6].

6.3 Advanced Design Requirements for Display Backlighting

Together with the energy consumption, the ability to reproduce the real colours of objects is one of the most desired properties for a liquid crystal display. These displays rely on a backlight which is a white light source that spans three colours, i.e., blue, green, and red. This white light passes through polarizers and then liquid crystals of each pixel that let the light pass through or block it. Next, the light reaches the blue, green, or red color filters that produce the desired color. The color span (also referred to as color gamut) of a display is governed by the purity of each blue, green, and red color components as a display can only produce colors made of combinations of individual color components. Thus, the purer individual color components are, the wider the color span is. Therefore, the narrower each color component is, the wider the color gamut of the display becomes. The narrow color components can be obtained by utilizing color filters having very narrow transmission spectra. Nevertheless, this is not a good idea unless the individual color components are narrow enough because a significant portion of the energy would otherwise be wasted. Therefore, the correct choice here is utilizing narrow-band emitters, e.g., nanocrystal quantum dots, colloidal quantum wells, perovskite nanocrystals, and individual LED chips.

To have an idea on the potential of narrow emitters for displays, in Fig. 6.5 we show the color gamut of the National Television System Committee (NTSC) standard (dashed triangle) together with the potential color gamut of displays using quantum dot integrated backlight sources [4]. The black lines in the graph indicate the chromaticity coordinates of the quantum dots with peak emission wavelengths

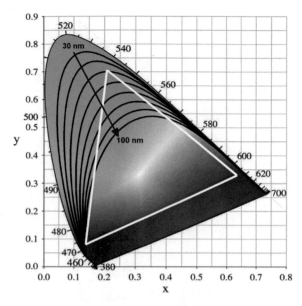

Fig. 6.5 NTSC color gamut in (x, y) color space (white triangle). The black lines indicate the colors that can be obtained using the nanocrystal quantum dots with different linewidths ranging from 30 to 100 nm. Reproduced from Ref. [4] © Science Wise Publishing & DeGruyter 2013

varying from 460 to 700 nm with linewidths between 30 and 100 nm. It is noticeable that carefully designed quantum dot backlight is able to produce a larger color gamut than the NTSC color gamut. This figure also tells us that the quantum dots should have a linewidth of 50 nm or narrower in order to satisfy the standard color gamut.

More detailed analyses on the requirements for a high-performance display employing nanocrystal quantum dots were carried by Luo et al. [9]. In this work, the performance of blue LED pumped color enrichment structure made of nanocrystal quantum dots was studied. The authors presented six different display backlighting designs that can achieve similar brightness values to those of the commercially available display technologies. These results summarized in Table 6.1 present designs for color gamuts of the designed displays in comparison with NTSC color gamut in (x, y) and (u′, v′) color spaces. Here, we see that if correctly designed, quantum dot based display backlighting can achieve 120 and 140% of the NTSC color gamut according to (x, y) and (u′, v′) color spaces, respectively. To achieve a broad color gamut, the emission peaks of blue and red color components have to be around 445 and 635 nm, respectively. The emission peak of the green color component seems to dictate the observed performance differences between two different color spaces. Another interesting feature is that the emission linewidth is not a strict requirement for the blue component, whereas narrow emitters definitely seem to be favored for the red and green color components.

At this point, it is worth commenting on the choice of color spaces for optimizing the display designs. If one wants to increase the number of colors that a display

Table 6.1 Optimal designs for nanocrystal quantum dot integrated displays based on (x, y) and (u′, v′) color spaces [10]

	Design #1	Design #2	Design #3	Design #4	Design #5	Design #6
λ_{blue} (nm)	454.1	452.0	450.1	449.7	447.6	443.2
λ_{green} (nm)	548.0	542.7	535.6	529.8	523.5	547.1
λ_{red} (nm)	606.5	611.3	615.0	621.8	634.8	635.7
$\Delta\lambda_{blue}$ (nm)	20.1	20.6	20.1	20.9	20.0	20.0
$\Delta\lambda_{green}$ (nm)	30.1	30.8	30.4	30.4	30.0	30.0
$\Delta\lambda_{red}$ (nm)	30.5	30.3	30.3	30.6	30.0	30.0
α_{blue} (%)	48.2	47.1	46.1	42.8	37.3	44.8
α_{green} (%)	30.8	29.8	28.8	29.3	27.8	30.2
α_{red} (%)	20.9	23.1	25.1	27.9	34.9	24.9
Color Gamut/NTSC Gamut (x, y) color space)	0.80	0.90	1.00	1.10	1.20	1.01
Color Gamut/NTSC Gamut (u′, v′) color space)	1.00	1.09	1.16	1.23	1.31	1.40

λ stands for the peak emission wavelength
$\Delta\lambda$ is the full-width at half-maximum, and
α is the relative amplitude of a particular color component

can show, obviously the color gamut has to be broadened. Nevertheless, to make a fair comparison but more importantly in order to have more meaningful and correct results, a uniform color gamut has to be used. This necessity becomes obvious when we take a look at the CIE 1931 (x, y) color space. In this system, the green colors [sparsely spaced on (x, y)] span a huge area whereas the blue and especially red colors are confined to smaller regions. An optimized design based on this system, therefore, will naturally focus on narrow green emitters and mostly disregard the spectral characteristics of blue and red emitters. On the other hand, the effects of each color components seem to be more balanced in uniform color spaces. However, in this case, we strongly urge the designers to employ modern uniform color spaces such as L*a*b* rather than old ones such as (u′, v′). As it is usually the case in color science, the industry is hardly adapted to these facts related to colorimetry and still tends to utilize old systems in their designs.

References

1. Erdem T, Nizamoglu S, Sun XW, Demir HV (2010) A photometric investigation of ultra-efficient LEDs with high color rendering index and high luminous efficacy employing nanocrystal quantum dot luminophores. Opt Express 18(1):340–347
2. Erdem T, Nizamoglu S, Demir HV (2012) Computational study of power conversion and luminous efficiency performance for semiconductor quantum dot nanophosphors on light-emitting diodes. Opt Express 20(3):3275–3295
3. Phillips JM et al (2007) Research challenges to ultra-efficient inorganic solid-state lighting. Laser Photonics Rev 1(4):307
4. Erdem T, Demir HV (2013) Color science of nanocrystal quantum dots for lighting and displays. Nanophotonics 2(1):57–81
5. Narukawa Y, Ichikawa M, Sanga D, Sano M, Mukai T (2010) White light emitting diodes with super-high luminous efficacy. J Phys D Appl Phys 43(35):354002
6. Erdem T, Kelestemur Y, Soran-Erdem Z, Ji Y, Demir HV (2014) Energy-saving quality road lighting with colloidal quantum dot nanophosphors. Nanophotonics 3(6):373–381
7. "British Standard BS 5489-1:2003—Code of practice for the design of road lighting," 2003
8. "American National Standard for electric lamps—Specifications for the chromaticity of solid state lighting products." American National Standards Institute, 2011
9. Luo Z, Chen Y, Wu S-T (2013) Wide color gamut LCD with a quantum dot backlight. Opt Express 21(22):26269–26284

Chapter 7
Future Outlook

Abstract In this final Chapter, we present a future perspective for the light source design. We discuss the existing problems and briefly introduce new material systems and device architectures that can overcome the current issues.

Keywords Narrow-band emitters · Lasers · Cd-free nanocrystals · Nanoplatelets · Perovskites

The performance metrics of high-quality lighting are numerous and their relations with each other are far from trivial. Therefore, defining guidelines for realizing high-quality lighting are challenging; nevertheless, in the literature related problems have been addressed, and important points are identified as also explained in this brief. A similar situation exists for the displays where the main requirement is broadening the color gamut, which was also addressed in the literature and covered in this brief. Clearly, the key point for both applications is the use of narrow-band emitters such as nanocrystal quantum dots. This is because these emitters enable precise control over the photometric characteristics of the light source leading to high-quality and high-efficiency lighting. Although important problems related to their integration into light sources have been resolved, there are still existing issues to address and there is room for further development.

Among those problems, the most well-known issue is the cadmium content of the efficient nanocrystals. To avoid the environmental risks of these issues and compounds, either a proper recycling method has to be developed or proper cadmium-free alternatives have to be synthesized. The existing inorganic cadmium-free nanocrystals possess significantly lower efficiencies compared to their cadmium-containing counterparts. Furthermore, their emission spectra are significantly broader. Therefore, even if we can increase their efficiencies using methods such as incorporation into macrocrystals and near-field interactions, their emission spectra should be definitely narrowed. Otherwise, high-quality lighting cannot be realized using these materials. As an alternative, developing cadmium-free colloidal quantum wells (for example InP) step forward if they are made efficiently emitting. In addition to inorganic nano-emitters, the use of organic nanoparticles may also be utilized for solid-state lighting. However, their emission spectra have to be narrowed before they can be employed ubiquitously in solid-state lighting.

© The Author(s), under exclusive licence to Springer Nature Singapore Pte Ltd. 2019   45
T. Erdem and H. V. Demir, *Color Science and Photometry for Lighting with LEDs and Semiconductor Nanocrystals*, Nanoscience and Nanotechnology,
https://doi.org/10.1007/978-981-13-5886-9_7

At this point, it is worth mentioning about a new material system that is very promising for solid-state lighting owing to their very narrow emission bands. These are colloidal nanoplatelets that are essentially very thin atomically flat semiconductors (having a few monolayers of crystal structure) with larger sizes in lateral dimensions [1, 2]. Their super small thickness making these particles colloidal analogs to quantum wells enables very narrow emission linewidths on the order of 7–8 nm at room temperature. Currently, the main difficulty of using these materials is the stability issues in their solid-films. Once these are resolved, these nanoplatelets can help achieving really high-quality indoor and outdoor lighting as well as displays with extremely broad color gamuts.

The main problem of color conversion is the unavoidable energy loss because of the absorption of a high-power photon but the emission of lower energy photons. Even if the quantum efficiency of the color convertors is unity, this energy loss cannot be eliminated. The solution to this issue is developing efficient single-color light sources and using them together to obtain white light. As we have explained in Chap. 5, the reason that currently color conversion based approaches are utilized is the lack of high-efficiency green and yellow LED chips. However, we know that both the industry and academia are putting a lot of effort to solve this problem. Furthermore, nanocrystal quantum dot LEDs that are electrically driven as opposed to color conversion are also very promising. These devices achieved external quantum efficiencies of 20% [3, 4], bringing them closer to be applied in displays and general lighting applications. Once this problem is solved, we believe that using combinations of single-color LED chips will form the essence of solid-state lighting.

When it comes to narrow-band emitters, one should also recall the lasers and discuss their performance for solid-state lighting applications. As these devices have linewidths of ca. 1–2 nm, using them in displays will tremendously improve the color gamut. Although it may seem not suitable, correct combinations of lasers are also shown to be very promising for indoor lighting application [5]. However, the green gap problem still remains as a challenge; but once it is solved, utilizing lasers can provide very high power efficiencies as well as high colorimetric and photometric quality. Therefore, it would not be wrong to express that lasers can be the future for general lighting and display applications.

References

1. Tessier MD, Mahler B, Nadal B, Heuclin H, Pedetti S, Dubertret B (2013) Spectroscopy of colloidal semiconductor core/shell nanoplatelets with high quantum yield. Nano Lett 13(7):3321–3328
2. Ithurria S, Tessier MD, Mahler B, Lobo RPSM, Dubertret B, Efros AL (2011) Colloidal nanoplatelets with two-dimensional electronic structure. Nat Mater 10(12):936–941
3. Mashford B, Stevenson M, Popovic Z (2013) High-efficiency quantum-dot light-emitting devices with enhanced charge injection. Nat Photonics 7(4):407–412

4. Erdem T, Demir HV (2016) Colloidal nanocrystals for quality lighting and displays: milestones and recent developments. Nanophotonics 5(1):74–95
5. Neumann A, Wierer JJ, Davis W, Ohno Y, Brueck SRJ, Tsao JY (2011) Four-color laser white illuminant demonstrating high color-rendering quality. Opt express 19(104):982–990

Appendix A
Tables of Colorimetric and Photometric Data

See Tables A.1, A.2, A.3, A.4 and A.5.

Table A.1 Photopic ($V(\lambda)$) and scotopic ($V'(\lambda)$) eye sensitivity functions

λ (nm)	$V(\lambda)$	$V'(\lambda)$
360	0.00000	0.00000
365	0.00000	0.00000
370	0.00000	0.00000
375	0.00000	0.00000
380	0.00020	0.00059
385	0.00040	0.00111
390	0.00080	0.00221
395	0.00155	0.00453
400	0.00280	0.00929
405	0.00466	0.01852
410	0.00740	0.03484
415	0.01178	0.06040
420	0.01750	0.09660
425	0.02268	0.14360
430	0.02730	0.19980
435	0.03258	0.26250
440	0.03790	0.32810
445	0.04239	0.39310
450	0.04680	0.45500
455	0.05212	0.51300
460	0.06000	0.56700
465	0.07294	0.62000
470	0.09098	0.67600

(continued)

© The Author(s), under exclusive licence to Springer Nature Singapore Pte Ltd. 2019
T. Erdem and H. V. Demir, *Color Science and Photometry for Lighting with LEDs and Semiconductor Nanocrystals*, Nanoscience and Nanotechnology,
https://doi.org/10.1007/978-981-13-5886-9

Table A.1 (continued)

λ (nm)	V(λ)	V'(λ)
475	0.11284	0.73400
480	0.13902	0.79300
485	0.16987	0.85100
490	0.20802	0.90400
495	0.25808	0.94900
500	0.32300	0.98200
505	0.40540	0.99800
510	0.50300	0.99700
515	0.60811	0.97500
520	0.71000	0.93500
525	0.79510	0.88000
530	0.86200	0.81100
535	0.91505	0.73300
540	0.95400	0.65000
545	0.98004	0.56400
550	0.99495	0.48100
555	1.00010	0.40200
560	0.99500	0.32880
565	0.97875	0.26390
570	0.95200	0.20760
575	0.91558	0.16020
580	0.87000	0.12120
585	0.81623	0.08990
590	0.75700	0.06550
595	0.69483	0.04690
600	0.63100	0.03315
605	0.56654	0.02312
610	0.50300	0.01593
615	0.44172	0.01088
620	0.38100	0.00737
625	0.32052	0.00497
630	0.26500	0.00334
635	0.21702	0.00224
640	0.17500	0.00150
645	0.13812	0.00101
650	0.10700	0.00068
655	0.08165	0.00046
660	0.06100	0.00031
665	0.04433	0.00021
670	0.03200	0.00015

(continued)

Table A.1 (continued)

λ (nm)	V(λ)	V'(λ)
675	0.02345	0.00010
680	0.01700	0.00007
685	0.01187	0.00005
690	0.00821	0.00004
695	0.00577	0.00003
700	0.00410	0.00002
705	0.00293	0.00001
710	0.00209	0.00001
715	0.00148	0.00001
720	0.00105	0.00000
725	0.00074	0.00000
730	0.00052	0.00000
735	0.00036	0.00000
740	0.00025	0.00000
745	0.00017	0.00000
750	0.00012	0.00000
755	0.00008	0.00000
760	0.00006	0.00000
765	0.00004	0.00000
770	0.00003	0.00000
775	0.00002	0.00000
780	0.00001	0.00000
785	0.00001	0.00000
790	0.00001	0.00000
795	0.00001	0.00000
800	0.00000	0.00000
805	0.00000	0.00000
810	0.00000	0.00000
815	0.00000	0.00000
820	0.00000	0.00000
825	0.00000	0.00000
830	0.00000	0.00000

Table A.2 Color-matching functions

λ (nm)	$\bar{x}$	$\bar{y}$	$\bar{z}$
360	0.00013	0.00000	0.00061
365	0.00023	0.00001	0.00109
370	0.00042	0.00001	0.00195
375	0.00074	0.00002	0.00349
380	0.00137	0.00004	0.00065
385	0.00224	0.00006	0.01055
390	0.00424	0.00012	0.02005
395	0.00765	0.00022	0.03621
400	0.01431	0.00040	0.06785
405	0.02319	0.00064	0.11020
410	0.04351	0.00121	0.20740
415	0.07763	0.00218	0.37130
420	0.13438	0.00400	0.64560
425	0.21477	0.00730	1.03905
430	0.28390	0.01160	1.38560
435	0.32850	0.01684	1.62296
440	0.34828	0.02300	1.74706
445	0.34806	0.02980	1.78260
450	0.33620	0.03800	1.77211
455	0.31870	0.04800	1.74410
460	0.29080	0.06000	1.66920
465	0.25110	0.07390	1.52810
470	0.19536	0.09098	1.28764
475	0.14210	0.11260	1.04190
480	0.09564	0.13902	0.81295
485	0.05795	0.16930	0.61620
490	0.03201	0.20802	0.46518
495	0.01470	0.25860	0.35330
500	0.00490	0.32300	0.27200
505	0.00240	0.40730	0.21230
510	0.00930	0.50300	0.15820
515	0.02910	0.60820	0.11170
520	0.06327	0.71000	0.07825
525	0.10960	0.79320	0.05725
530	0.16550	0.86200	0.04216
535	0.22575	0.91485	0.02984
540	0.29040	0.95400	0.02030
545	0.35970	0.98030	0.01340
550	0.43345	0.99495	0.00875
555	0.51205	1.00000	0.00575

(continued)

Table A.2 (continued)

λ (nm)	$\bar{x}$	$\bar{y}$	$\bar{z}$
560	0.59450	0.99500	0.00390
565	0.67840	0.97860	0.00275
570	0.76210	0.95200	0.00210
575	0.84250	0.91540	0.00180
580	0.91630	0.87000	0.00165
585	0.97860	0.81630	0.00140
590	1.02630	0.75700	0.00110
595	1.05670	0.69490	0.00100
600	1.06220	0.63100	0.00080
605	1.04560	0.56680	0.00060
610	1.00260	0.50300	0.00034
615	0.93840	0.44120	0.00024
620	0.85450	0.38100	0.00019
625	0.75140	0.32100	0.00010
630	0.64240	0.26500	0.00050
635	0.54190	0.21700	0.00030
640	0.44790	0.17500	0.00020
645	0.36080	0.13800	0.00010
650	0.28350	0.10700	0.00000
655	0.21870	0.08160	0.00000
660	0.16490	0.06100	0.00000
665	0.12120	0.04458	0.00000
670	0.08740	0.03200	0.00000
675	0.06360	0.02320	0.00000
680	0.04677	0.01700	0.00000
685	0.03290	0.01192	0.00000
690	0.02270	0.00821	0.00000
695	0.01584	0.00572	0.00000
700	0.01136	0.00410	0.00000
705	0.00811	0.00293	0.00000
710	0.00579	0.00209	0.00000
715	0.00411	0.00148	0.00000
720	0.00290	0.00105	0.00000
725	0.00205	0.00074	0.00000
730	0.00144	0.00052	0.00000
735	0.00100	0.00036	0.00000
740	0.00069	0.00025	0.00000
745	0.00048	0.00017	0.00000
750	0.00033	0.00012	0.00000
755	0.00024	0.00008	0.00000

(continued)

Table A.2 (continued)

λ (nm)	$\bar{x}$	$\bar{y}$	$\bar{z}$
760	0.00017	0.00006	0.00000
765	0.00018	0.00004	0.00000
770	0.00008	0.00003	0.00000
775	0.00006	0.00002	0.00000
780	0.00004	0.00002	0.00000
785	0.00003	0.00001	0.00000
790	0.00002	0.00001	0.00000
795	0.00001	0.00001	0.00000
800	0.00001	0.00000	0.00000
805	0.00001	0.00000	0.00000
810	0.00001	0.00000	0.00000
815	0.00000	0.00000	0.00000
820	0.00000	0.00000	0.00000
825	0.00000	0.00000	0.00000
830	0.00000	0.00000	0.00000

Table A.3 Relative spectral power distributions of CIE standard illuminants A and D65

λ (nm)	D65	Illuminant A
360	0.34000	6.14462
365	0.36000	6.94720
370	0.38000	7.82135
375	0.40000	8.76980
380	0.42420	9.79510
385	0.44403	10.89960
390	0.46386	12.08530
395	0.58315	13.35430
400	0.70243	14.70800
405	0.73949	16.14800
410	0.77654	17.67530
415	0.78480	19.29070
420	0.79306	20.99500
425	0.76441	22.78830
430	0.73577	24.67090
435	0.81294	26.64250
440	0.89010	28.70270
445	0.94164	30.85080
450	0.99318	33.08590
455	0.99659	35.40680
460	1.00000	37.81210
465	0.98747	40.30020
470	0.97495	42.86930
475	0.97946	45.51740
480	0.98397	48.24230
485	0.95378	51.04180
490	0.92360	53.91320
495	0.92590	56.85390
500	0.92821	59.86110
505	0.92162	62.93200
510	0.91503	66.06350
515	0.90225	69.25250
520	0.88947	72.49590
525	0.90177	75.79030
530	0.91407	79.13260
535	0.90014	82.51930
540	0.88620	85.94700
545	0.88467	89.41240
550	0.88315	92.91200
555	0.86598	96.44230

(continued)

Table A.3 (continued)

λ (nm)	D65	Illuminant A
560	0.84881	100.00000
565	0.83325	103.58200
570	0.81769	107.18400
575	0.81538	110.80300
580	0.81306	114.43600
585	0.78292	118.08000
590	0.75277	121.73100
595	0.75838	125.38600
600	0.76398	129.04300
605	0.76225	132.69700
610	0.76053	136.34600
615	0.75246	139.98800
620	0.74440	143.61800
625	0.72568	147.23500
630	0.70696	150.83600
635	0.70870	154.41800
640	0.71045	157.97900
645	0.69486	161.51600
650	0.67928	165.02800
655	0.68007	168.51000
660	0.68087	171.96300
665	0.68963	175.38300
670	0.69838	178.76900
675	0.68143	182.11800
680	0.66448	185.42900
685	0.62814	188.70100
690	0.59180	191.93100
695	0.59981	195.11800
700	0.60783	198.26100
705	0.61945	201.35900
710	0.63108	204.40900
715	0.57699	207.41100
720	0.52290	210.36500
725	0.55805	213.26800
730	0.59320	216.12000
735	0.61527	218.92000
740	0.63735	221.66700
745	0.58856	224.36100
750	0.53978	227.00000
755	0.46689	229.58500

(continued)

Table A.3 (continued)

λ (nm)	D65	Illuminant A
760	0.39400	232.11500
765	0.48053	234.58900
770	0.56705	237.00800
775	0.55253	239.37000
780	0.53800	241.67500
785	0.53000	243.92400
790	0.52000	246.11600
795	0.51000	248.25100
800	0.50000	250.32900
805	0.49000	252.35000
810	0.48000	254.31400
815	0.47000	256.22100
820	0.46000	258.07100
825	0.45000	259.86500
830	0.44000	261.60200

Table A.4 Spectral reflectivities of test color samples that are used to calculate the color rendering index (CRI)

λ (nm)	TCS 01	TCS 02	TCS 03	TCS 04	TCS 05	TCS 06	TCS 07	TCS 08	TCS 09	TCS 10	TCS 11	TCS 12	TCS 13	TCS 14
360	0.116	0.053	0.058	0.057	0.143	0.079	0.150	0.075	0.069	0.042	0.074	0.189	0.071	0.036
365	0.136	0.055	0.059	0.059	0.187	0.081	0.177	0.078	0.072	0.043	0.079	0.175	0.076	0.036
370	0.159	0.059	0.061	0.062	0.233	0.089	0.218	0.084	0.073	0.045	0.086	0.158	0.082	0.036
375	0.190	0.064	0.063	0.067	0.269	0.113	0.293	0.090	0.070	0.047	0.098	0.139	0.090	0.036
380	0.219	0.070	0.065	0.074	0.295	0.151	0.378	0.104	0.066	0.050	0.111	0.120	0.104	0.036
385	0.239	0.079	0.068	0.083	0.306	0.203	0.459	0.129	0.062	0.054	0.121	0.103	0.127	0.036
390	0.252	0.089	0.070	0.093	0.310	0.265	0.524	0.170	0.058	0.059	0.127	0.090	0.161	0.037
395	0.256	0.101	0.072	0.105	0.312	0.339	0.546	0.240	0.055	0.063	0.129	0.082	0.211	0.038
400	0.256	0.111	0.073	0.116	0.313	0.410	0.551	0.319	0.052	0.066	0.127	0.076	0.264	0.039
405	0.254	0.116	0.073	0.121	0.315	0.464	0.555	0.416	0.052	0.067	0.121	0.068	0.313	0.039
410	0.252	0.118	0.074	0.124	0.319	0.492	0.559	0.462	0.051	0.068	0.116	0.064	0.341	0.040
415	0.248	0.120	0.074	0.126	0.322	0.508	0.560	0.482	0.050	0.069	0.112	0.065	0.352	0.041
420	0.244	0.121	0.074	0.128	0.326	0.517	0.561	0.490	0.050	0.069	0.108	0.075	0.359	0.042
425	0.240	0.122	0.073	0.131	0.330	0.524	0.558	0.488	0.049	0.070	0.105	0.093	0.361	0.042
430	0.237	0.122	0.073	0.135	0.334	0.531	0.556	0.482	0.048	0.072	0.104	0.123	0.364	0.043
435	0.232	0.122	0.073	0.139	0.339	0.538	0.551	0.473	0.047	0.073	0.104	0.160	0.365	0.044
440	0.230	0.123	0.073	0.144	0.346	0.544	0.544	0.462	0.046	0.076	0.105	0.207	0.367	0.044
445	0.226	0.124	0.073	0.151	0.352	0.551	0.535	0.450	0.044	0.078	0.106	0.256	0.369	0.045
450	0.225	0.127	0.074	0.161	0.360	0.556	0.522	0.439	0.042	0.083	0.110	0.300	0.372	0.045
455	0.222	0.128	0.075	0.172	0.369	0.556	0.506	0.426	0.041	0.088	0.115	0.331	0.374	0.046
460	0.220	0.131	0.077	0.186	0.381	0.554	0.488	0.413	0.038	0.095	0.123	0.346	0.376	0.047
465	0.218	0.134	0.080	0.205	0.394	0.549	0.469	0.397	0.035	0.103	0.134	0.347	0.379	0.048
470	0.216	0.138	0.085	0.229	0.403	0.541	0.448	0.382	0.033	0.113	0.148	0.341	0.384	0.050

(continued)

Table A.4 (continued)

λ (nm)	TCS 01	TCS 02	TCS 03	TCS 04	TCS 05	TCS 06	TCS 07	TCS 08	TCS 09	TCS 10	TCS 11	TCS 12	TCS 13	TCS 14
475	0.214	0.143	0.094	0.254	0.410	0.531	0.429	0.366	0.031	0.125	0.167	0.328	0.389	0.052
480	0.214	0.150	0.109	0.281	0.415	0.519	0.408	0.352	0.030	0.142	0.192	0.307	0.397	0.055
485	0.214	0.159	0.126	0.308	0.418	0.504	0.385	0.337	0.029	0.162	0.219	0.282	0.405	0.057
490	0.216	0.174	0.148	0.332	0.419	0.488	0.363	0.325	0.028	0.189	0.252	0.257	0.416	0.062
495	0.218	0.190	0.172	0.352	0.417	0.469	0.341	0.310	0.028	0.219	0.291	0.230	0.429	0.067
500	0.223	0.207	0.198	0.370	0.413	0.450	0.324	0.299	0.028	0.262	0.325	0.204	0.443	0.075
505	0.225	0.225	0.221	0.383	0.409	0.431	0.311	0.289	0.029	0.305	0.347	0.178	0.454	0.083
510	0.226	0.242	0.241	0.390	0.403	0.414	0.301	0.283	0.030	0.365	0.356	0.154	0.461	0.092
515	0.226	0.253	0.260	0.394	0.396	0.395	0.291	0.276	0.030	0.416	0.353	0.129	0.466	0.100
520	0.225	0.260	0.278	0.395	0.389	0.377	0.283	0.270	0.031	0.465	0.346	0.109	0.469	0.108
525	0.225	0.264	0.302	0.392	0.381	0.358	0.273	0.262	0.031	0.509	0.333	0.090	0.471	0.121
530	0.227	0.267	0.339	0.385	0.372	0.341	0.265	0.256	0.032	0.546	0.314	0.075	0.474	0.133
535	0.230	0.269	0.370	0.377	0.363	0.325	0.260	0.251	0.032	0.581	0.294	0.062	0.476	0.142
540	0.236	0.272	0.392	0.367	0.353	0.309	0.257	0.250	0.033	0.610	0.271	0.051	0.483	0.150
545	0.245	0.276	0.399	0.354	0.342	0.293	0.257	0.251	0.034	0.634	0.248	0.041	0.490	0.154
550	0.253	0.282	0.400	0.341	0.331	0.279	0.259	0.254	0.035	0.653	0.227	0.035	0.506	0.155
555	0.262	0.289	0.393	0.327	0.320	0.265	0.260	0.258	0.037	0.666	0.206	0.029	0.526	0.152
560	0.272	0.299	0.380	0.312	0.308	0.253	0.260	0.264	0.041	0.678	0.188	0.025	0.553	0.147
565	0.283	0.309	0.365	0.296	0.296	0.241	0.258	0.269	0.044	0.687	0.170	0.022	0.582	0.140
570	0.298	0.322	0.349	0.280	0.284	0.234	0.256	0.272	0.048	0.693	0.153	0.019	0.618	0.133
575	0.318	0.329	0.332	0.263	0.271	0.227	0.254	0.274	0.052	0.698	0.138	0.017	0.651	0.125
580	0.341	0.335	0.315	0.247	0.260	0.225	0.254	0.278	0.060	0.701	0.125	0.017	0.680	0.118
585	0.367	0.339	0.299	0.229	0.247	0.222	0.259	0.284	0.076	0.704	0.114	0.017	0.701	0.112

(continued)

Table A.4 (continued)

λ (nm)	TCS 01	TCS 02	TCS 03	TCS 04	TCS 05	TCS 06	TCS 07	TCS 08	TCS 09	TCS 10	TCS 11	TCS 12	TCS 13	TCS 14
590	0.390	0.341	0.285	0.214	0.232	0.221	0.270	0.295	0.102	0.705	0.106	0.016	0.717	0.106
595	0.409	0.341	0.272	0.198	0.220	0.220	0.284	0.316	0.136	0.705	0.100	0.016	0.729	0.101
600	0.424	0.342	0.264	0.185	0.210	0.220	0.302	0.348	0.190	0.706	0.096	0.016	0.736	0.098
605	0.435	0.342	0.257	0.175	0.200	0.220	0.324	0.384	0.256	0.707	0.092	0.016	0.742	0.095
610	0.442	0.342	0.252	0.169	0.194	0.220	0.344	0.434	0.336	0.707	0.090	0.016	0.745	0.093
615	0.448	0.341	0.247	0.164	0.189	0.220	0.362	0.482	0.418	0.707	0.087	0.016	0.747	0.090
620	0.450	0.341	0.241	0.160	0.185	0.223	0.377	0.528	0.505	0.708	0.085	0.016	0.748	0.089
625	0.451	0.339	0.235	0.156	0.183	0.227	0.389	0.568	0.581	0.708	0.082	0.016	0.748	0.087
630	0.451	0.339	0.229	0.154	0.180	0.233	0.400	0.604	0.641	0.710	0.080	0.018	0.748	0.086
635	0.451	0.338	0.224	0.152	0.177	0.239	0.410	0.629	0.682	0.711	0.079	0.018	0.748	0.085
640	0.451	0.338	0.220	0.151	0.176	0.244	0.420	0.648	0.717	0.712	0.078	0.018	0.748	0.084
645	0.451	0.337	0.217	0.149	0.175	0.251	0.429	0.663	0.740	0.714	0.078	0.018	0.748	0.084
650	0.450	0.336	0.216	0.148	0.175	0.258	0.438	0.676	0.758	0.716	0.078	0.019	0.748	0.084
655	0.450	0.335	0.216	0.148	0.175	0.263	0.445	0.685	0.770	0.718	0.078	0.020	0.748	0.084
660	0.451	0.334	0.219	0.148	0.175	0.268	0.452	0.693	0.781	0.720	0.081	0.023	0.747	0.085
665	0.451	0.332	0.224	0.149	0.177	0.273	0.457	0.700	0.790	0.722	0.083	0.024	0.747	0.087
670	0.453	0.332	0.230	0.151	0.180	0.278	0.462	0.705	0.797	0.725	0.088	0.026	0.747	0.092
675	0.454	0.331	0.238	0.154	0.183	0.281	0.466	0.709	0.803	0.729	0.093	0.030	0.747	0.096
680	0.455	0.331	0.251	0.158	0.186	0.283	0.468	0.712	0.809	0.731	0.102	0.035	0.747	0.102
685	0.457	0.330	0.269	0.162	0.189	0.286	0.470	0.715	0.814	0.735	0.112	0.043	0.747	0.110
690	0.458	0.329	0.288	0.165	0.192	0.291	0.473	0.717	0.819	0.739	0.125	0.056	0.747	0.123
695	0.460	0.328	0.312	0.168	0.195	0.296	0.477	0.719	0.824	0.742	0.141	0.074	0.746	0.137
700	0.462	0.328	0.340	0.170	0.199	0.302	0.483	0.721	0.828	0.746	0.161	0.097	0.746	0.152

(continued)

Table A.4 (continued)

λ (nm)	TCS 01	TCS 02	TCS 03	TCS 04	TCS 05	TCS 06	TCS 07	TCS 08	TCS 09	TCS 10	TCS 11	TCS 12	TCS 13	TCS 14
705	0.463	0.327	0.366	0.171	0.200	0.313	0.489	0.720	0.830	0.748	0.182	0.128	0.746	0.169
710	0.464	0.326	0.390	0.170	0.199	0.325	0.496	0.719	0.831	0.749	0.203	0.166	0.745	0.188
715	0.465	0.325	0.412	0.168	0.198	0.338	0.503	0.722	0.833	0.751	0.223	0.210	0.744	0.207
720	0.466	0.324	0.431	0.166	0.196	0.351	0.511	0.725	0.835	0.753	0.242	0.257	0.743	0.226
725	0.466	0.324	0.447	0.164	0.195	0.364	0.518	0.727	0.836	0.754	0.257	0.305	0.744	0.243
730	0.466	0.324	0.460	0.164	0.195	0.376	0.525	0.729	0.836	0.755	0.270	0.354	0.745	0.260
735	0.466	0.323	0.472	0.165	0.196	0.389	0.532	0.730	0.837	0.755	0.282	0.401	0.748	0.277
740	0.467	0.322	0.481	0.168	0.197	0.401	0.539	0.730	0.838	0.755	0.292	0.446	0.750	0.294
745	0.467	0.321	0.488	0.172	0.200	0.413	0.546	0.730	0.839	0.755	0.302	0.485	0.750	0.310
750	0.467	0.320	0.493	0.177	0.203	0.425	0.553	0.730	0.839	0.756	0.310	0.520	0.749	0.325
755	0.467	0.318	0.497	0.181	0.205	0.436	0.559	0.730	0.839	0.757	0.314	0.551	0.748	0.339
760	0.467	0.316	0.500	0.185	0.208	0.447	0.565	0.730	0.839	0.758	0.317	0.577	0.748	0.353
765	0.467	0.315	0.502	0.189	0.212	0.458	0.570	0.730	0.839	0.759	0.323	0.599	0.747	0.366
770	0.467	0.315	0.505	0.192	0.215	0.469	0.575	0.730	0.839	0.759	0.330	0.618	0.747	0.379
775	0.467	0.314	0.510	0.194	0.217	0.477	0.578	0.730	0.839	0.759	0.334	0.633	0.747	0.390
780	0.467	0.314	0.516	0.197	0.219	0.485	0.581	0.730	0.839	0.759	0.338	0.645	0.747	0.399
785	0.467	0.313	0.520	0.200	0.222	0.493	0.583	0.730	0.839	0.759	0.343	0.656	0.746	0.408
790	0.467	0.313	0.524	0.204	0.226	0.500	0.585	0.731	0.839	0.759	0.348	0.666	0.746	0.416
795	0.466	0.312	0.527	0.210	0.231	0.506	0.587	0.731	0.839	0.759	0.353	0.674	0.746	0.422
800	0.466	0.312	0.531	0.218	0.237	0.512	0.588	0.731	0.839	0.759	0.359	0.680	0.746	0.428
805	0.466	0.311	0.535	0.225	0.243	0.517	0.589	0.731	0.839	0.759	0.365	0.686	0.745	0.434
810	0.466	0.311	0.539	0.233	0.249	0.521	0.590	0.731	0.838	0.758	0.372	0.691	0.745	0.439
815	0.466	0.311	0.544	0.243	0.257	0.525	0.590	0.731	0.837	0.757	0.380	0.694	0.745	0.444

(continued)

Table A.4 (continued)

λ (nm)	TCS 01	TCS 02	TCS 03	TCS 04	TCS 05	TCS 06	TCS 07	TCS 08	TCS 09	TCS 10	TCS 11	TCS 12	TCS 13	TCS 14
820	0.465	0.311	0.548	0.254	0.265	0.529	0.590	0.731	0.837	0.757	0.388	0.697	0.745	0.448
825	0.464	0.311	0.552	0.264	0.273	0.532	0.591	0.731	0.836	0.756	0.396	0.700	0.745	0.451
830	0.464	0.310	0.555	0.274	0.280	0.535	0.592	0.731	0.836	0.756	0.403	0.702	0.745	0.454

Table A.5 Spectral reflectivities of test color samples that are used to calculate the color quality scale (CQS)

λ (nm)	Sample 1	Sample 2	Sample 3	Sample 4	Sample 5	Sample 6	Sample 7	Sample 8	Sample 9	Sample 10	Sample 11	Sample 12	Sample 13	Sample 14	Sample 15
360	0.1086	0.1053	0.0858	0.0790	0.1167	0.0872	0.0726	0.0652	0.0643	0.0540	0.0482	0.0691	0.0829	0.0530	0.0908
365	0.1086	0.1053	0.0858	0.0790	0.1167	0.0872	0.0726	0.0652	0.0643	0.0540	0.0482	0.0691	0.0829	0.0530	0.0908
370	0.1086	0.1053	0.0858	0.0790	0.1167	0.0872	0.0726	0.0652	0.0643	0.0540	0.0482	0.0691	0.0829	0.0530	0.0908
375	0.1086	0.1053	0.0858	0.0790	0.1167	0.0872	0.0726	0.0652	0.0643	0.0540	0.0482	0.0691	0.0829	0.0530	0.0908
380	0.1086	0.1053	0.0858	0.0790	0.1167	0.0872	0.0726	0.0652	0.0643	0.0540	0.0482	0.0691	0.0829	0.0530	0.0908
385	0.1380	0.1323	0.0990	0.0984	0.1352	0.1001	0.0760	0.0657	0.0661	0.0489	0.0456	0.0692	0.0829	0.0507	0.1021
390	0.1729	0.1662	0.1204	0.1242	0.1674	0.1159	0.0789	0.0667	0.0702	0.0548	0.0478	0.0727	0.0866	0.0505	0.1130
395	0.2167	0.2113	0.1458	0.1595	0.2024	0.1339	0.0844	0.0691	0.0672	0.0550	0.0455	0.0756	0.0888	0.0502	0.1280
400	0.2539	0.2516	0.1696	0.1937	0.2298	0.1431	0.0864	0.0694	0.0715	0.0529	0.0484	0.0770	0.0884	0.0498	0.1359
405	0.2785	0.2806	0.1922	0.2215	0.2521	0.1516	0.0848	0.0709	0.0705	0.0521	0.0494	0.0806	0.0853	0.0489	0.1378
410	0.2853	0.2971	0.2101	0.2419	0.2635	0.1570	0.0861	0.0707	0.0727	0.0541	0.0456	0.0771	0.0868	0.0503	0.1363
415	0.2883	0.3042	0.2179	0.2488	0.2702	0.1608	0.0859	0.0691	0.0731	0.0548	0.0470	0.0742	0.0859	0.0492	0.1363
420	0.2860	0.3125	0.2233	0.2603	0.2758	0.1649	0.0868	0.0717	0.0745	0.0541	0.0473	0.0766	0.0828	0.0511	0.1354
425	0.2761	0.3183	0.2371	0.2776	0.2834	0.1678	0.0869	0.0692	0.0770	0.0531	0.0486	0.0733	0.0819	0.0509	0.1322
430	0.2674	0.3196	0.2499	0.2868	0.2934	0.1785	0.0882	0.0710	0.0756	0.0599	0.0501	0.0758	0.0822	0.0496	0.1294
435	0.2565	0.3261	0.2674	0.3107	0.3042	0.1829	0.0903	0.0717	0.0773	0.0569	0.0480	0.0768	0.0818	0.0494	0.1241
440	0.2422	0.3253	0.2949	0.3309	0.3201	0.1896	0.0924	0.0722	0.0786	0.0603	0.0490	0.0775	0.0822	0.0480	0.1209
445	0.2281	0.3193	0.3232	0.3515	0.3329	0.2032	0.0951	0.0737	0.0818	0.0643	0.0468	0.0754	0.0819	0.0487	0.1137
450	0.2140	0.3071	0.3435	0.3676	0.3511	0.2120	0.0969	0.0731	0.0861	0.0702	0.0471	0.0763	0.0807	0.0468	0.1117
455	0.2004	0.2961	0.3538	0.3819	0.3724	0.2294	0.1003	0.0777	0.0907	0.0715	0.0486	0.0763	0.0787	0.0443	0.1045
460	0.1854	0.2873	0.3602	0.4026	0.4027	0.2539	0.1083	0.0823	0.0981	0.0798	0.0517	0.0752	0.0832	0.0440	0.1006
465	0.1733	0.2729	0.3571	0.4189	0.4367	0.2869	0.1203	0.0917	0.1067	0.0860	0.0519	0.0782	0.0828	0.0427	0.0970
470	0.1602	0.2595	0.3511	0.4317	0.4625	0.3170	0.1383	0.1062	0.1152	0.0959	0.0479	0.0808	0.0810	0.0421	0.0908
475	0.1499	0.2395	0.3365	0.4363	0.4890	0.3570	0.1634	0.1285	0.1294	0.1088	0.0494	0.0778	0.0819	0.0414	0.0858

(continued)

Table A.5 (continued)

λ (nm)	Sample 1	Sample 2	Sample 3	Sample 4	Sample 5	Sample 6	Sample 7	Sample 8	Sample 9	Sample 10	Sample 11	Sample 12	Sample 13	Sample 14	Sample 15
480	0.1414	0.2194	0.3176	0.4356	0.5085	0.3994	0.1988	0.1598	0.1410	0.1218	0.0524	0.0788	0.0836	0.0408	0.0807
485	0.1288	0.1949	0.2956	0.4297	0.5181	0.4346	0.2376	0.1993	0.1531	0.1398	0.0527	0.0805	0.0802	0.0400	0.0752
490	0.1204	0.1732	0.2747	0.4199	0.5243	0.4615	0.2795	0.2445	0.1694	0.1626	0.0537	0.0809	0.0809	0.0392	0.0716
495	0.1104	0.1560	0.2506	0.4058	0.5179	0.4747	0.3275	0.2974	0.1919	0.1878	0.0577	0.0838	0.0838	0.0406	0.0688
500	0.1061	0.1436	0.2279	0.3882	0.5084	0.4754	0.3671	0.3462	0.2178	0.2302	0.0647	0.0922	0.0842	0.0388	0.0678
505	0.1018	0.1305	0.2055	0.3660	0.4904	0.4691	0.4030	0.3894	0.2560	0.2829	0.0737	0.1051	0.0865	0.0396	0.0639
510	0.0968	0.1174	0.1847	0.3433	0.4717	0.4556	0.4201	0.4180	0.3110	0.3455	0.0983	0.1230	0.0910	0.0397	0.0615
515	0.0941	0.1075	0.1592	0.3148	0.4467	0.4371	0.4257	0.4433	0.3789	0.4171	0.1396	0.1521	0.0920	0.0391	0.0586
520	0.0881	0.0991	0.1438	0.2890	0.4207	0.4154	0.4218	0.4548	0.4515	0.4871	0.1809	0.1728	0.0917	0.0405	0.0571
525	0.0842	0.0925	0.1244	0.2583	0.3931	0.3937	0.4090	0.4605	0.5285	0.5529	0.2280	0.1842	0.0917	0.0394	0.0527
530	0.0808	0.0916	0.1105	0.2340	0.3653	0.3737	0.3977	0.4647	0.5845	0.5955	0.2645	0.1897	0.0952	0.0401	0.0513
535	0.0779	0.0896	0.0959	0.2076	0.3363	0.3459	0.3769	0.4626	0.6261	0.6299	0.2963	0.1946	0.0983	0.0396	0.0537
540	0.0782	0.0897	0.0871	0.1839	0.3083	0.3203	0.3559	0.4604	0.6458	0.6552	0.3202	0.2037	0.1036	0.0396	0.0512
545	0.0773	0.0893	0.0790	0.1613	0.2808	0.2941	0.3312	0.4522	0.6547	0.6661	0.3545	0.2248	0.1150	0.0395	0.0530
550	0.0793	0.0891	0.0703	0.1434	0.2538	0.2715	0.3072	0.4444	0.6545	0.6752	0.3950	0.2675	0.1331	0.0399	0.0517
555	0.0790	0.0868	0.0652	0.1243	0.2260	0.2442	0.2803	0.4321	0.6473	0.6832	0.4353	0.3286	0.1646	0.0420	0.0511
560	0.0793	0.0820	0.0555	0.1044	0.2024	0.2205	0.2532	0.4149	0.6351	0.6851	0.4577	0.3895	0.2070	0.0410	0.0507
565	0.0806	0.0829	0.0579	0.0978	0.1865	0.1979	0.2313	0.4039	0.6252	0.6964	0.4904	0.4654	0.2754	0.0464	0.0549
570	0.0805	0.0854	0.0562	0.0910	0.1697	0.1800	0.2109	0.3879	0.6064	0.6966	0.5075	0.5188	0.3279	0.0500	0.0559
575	0.0793	0.0871	0.0548	0.0832	0.1592	0.1610	0.1897	0.3694	0.5924	0.7063	0.5193	0.5592	0.3819	0.0545	0.0627
580	0.0803	0.0922	0.0517	0.0771	0.1482	0.1463	0.1723	0.3526	0.5756	0.7104	0.5273	0.5909	0.4250	0.0620	0.0678
585	0.0815	0.0978	0.0544	0.0747	0.1393	0.1284	0.1528	0.3288	0.5549	0.7115	0.5359	0.6189	0.4690	0.0742	0.0810
590	0.0842	0.1037	0.0519	0.0726	0.1316	0.1172	0.1355	0.3080	0.5303	0.7145	0.5431	0.6343	0.5067	0.0937	0.1004
595	0.0912	0.1079	0.0520	0.0682	0.1217	0.1045	0.1196	0.2829	0.5002	0.7195	0.5449	0.6485	0.5443	0.1279	0.1268

(continued)

Table A.5 (continued)

λ (nm)	Sample 1	Sample 2	Sample 3	Sample 4	Sample 5	Sample 6	Sample 7	Sample 8	Sample 9	Sample 10	Sample 11	Sample 12	Sample 13	Sample 14	Sample 15
600	0.1035	0.1092	0.0541	0.0671	0.1182	0.0964	0.1050	0.2591	0.4793	0.7183	0.5493	0.6607	0.5721	0.1762	0.1595
605	0.1212	0.1088	0.0537	0.0660	0.1112	0.0903	0.0949	0.2388	0.4517	0.7208	0.5526	0.6648	0.5871	0.2449	0.2012
610	0.1455	0.1078	0.0545	0.0661	0.1071	0.0873	0.0868	0.2228	0.4340	0.7228	0.5561	0.6654	0.6073	0.3211	0.2452
615	0.1785	0.1026	0.0560	0.0660	0.1059	0.0846	0.0797	0.2109	0.4169	0.7274	0.5552	0.6721	0.6141	0.4050	0.2953
620	0.2107	0.0991	0.0560	0.0653	0.1044	0.0829	0.0783	0.2033	0.4060	0.7251	0.5573	0.6744	0.6170	0.4745	0.3439
625	0.2460	0.0995	0.0561	0.0644	0.1021	0.0814	0.0732	0.1963	0.3989	0.7274	0.5620	0.6723	0.6216	0.5335	0.3928
630	0.2791	0.1043	0.0578	0.0653	0.0991	0.0805	0.0737	0.1936	0.3945	0.7341	0.5607	0.6811	0.6272	0.5776	0.4336
635	0.3074	0.1101	0.0586	0.0669	0.1000	0.0803	0.0709	0.1887	0.3887	0.7358	0.5599	0.6792	0.6287	0.6094	0.4723
640	0.3330	0.1187	0.0573	0.0660	0.0980	0.0801	0.0703	0.1847	0.3805	0.7362	0.5632	0.6774	0.6276	0.6320	0.4996
645	0.3542	0.1311	0.0602	0.0677	0.0963	0.0776	0.0696	0.1804	0.3741	0.7354	0.5644	0.6796	0.6351	0.6495	0.5279
650	0.3745	0.1430	0.0604	0.0668	0.0997	0.0797	0.0673	0.1766	0.3700	0.7442	0.5680	0.6856	0.6362	0.6620	0.5428
655	0.3920	0.1583	0.0606	0.0693	0.0994	0.0801	0.0677	0.1734	0.3630	0.7438	0.5660	0.6853	0.6348	0.6743	0.5601
660	0.4052	0.1704	0.0606	0.0689	0.1022	0.0810	0.0682	0.1721	0.3640	0.7440	0.5709	0.6864	0.6418	0.6833	0.5736
665	0.4186	0.1846	0.0595	0.0676	0.1005	0.0819	0.0665	0.1720	0.3590	0.7436	0.5692	0.6879	0.6438	0.6895	0.5837
670	0.4281	0.1906	0.0609	0.0694	0.1044	0.0856	0.0691	0.1724	0.3648	0.7442	0.5657	0.6874	0.6378	0.6924	0.5890
675	0.4395	0.1983	0.0605	0.0687	0.1073	0.0913	0.0695	0.1757	0.3696	0.7489	0.5716	0.6871	0.6410	0.7030	0.5959
680	0.4440	0.1981	0.0602	0.0698	0.1069	0.0930	0.0723	0.1781	0.3734	0.7435	0.5729	0.6863	0.6460	0.7075	0.5983
685	0.4497	0.1963	0.0580	0.0679	0.1103	0.0958	0.0727	0.1829	0.3818	0.7460	0.5739	0.6890	0.6451	0.7112	0.6015
690	0.4555	0.2003	0.0587	0.0694	0.1104	0.1016	0.0757	0.1897	0.3884	0.7518	0.5714	0.6863	0.6432	0.7187	0.6054
695	0.4612	0.2034	0.0573	0.0675	0.1084	0.1044	0.0767	0.1949	0.3947	0.7550	0.5741	0.6893	0.6509	0.7214	0.6135
700	0.4663	0.2061	0.0606	0.0676	0.1092	0.1047	0.0810	0.2018	0.4011	0.7496	0.5774	0.6950	0.6517	0.7284	0.6200
705	0.4707	0.2120	0.0613	0.0662	0.1074	0.1062	0.0818	0.2051	0.4040	0.7548	0.5791	0.6941	0.6514	0.7327	0.6287
710	0.4783	0.2207	0.0618	0.0681	0.1059	0.1052	0.0837	0.2071	0.4072	0.7609	0.5801	0.6958	0.6567	0.7351	0.6405
715	0.4778	0.2257	0.0652	0.0706	0.1082	0.1029	0.0822	0.2066	0.4065	0.7580	0.5804	0.6950	0.6597	0.7374	0.6443

(continued)

Table A.5 (continued)

λ (nm)	Sample 1	Sample 2	Sample 3	Sample 4	Sample 5	Sample 6	Sample 7	Sample 8	Sample 9	Sample 10	Sample 11	Sample 12	Sample 13	Sample 14	Sample 15
720	0.4844	0.2335	0.0647	0.0728	0.1106	0.1025	0.0838	0.2032	0.4006	0.7574	0.5840	0.7008	0.6576	0.7410	0.6489
725	0.4877	0.2441	0.0684	0.0766	0.1129	0.1008	0.0847	0.1998	0.3983	0.7632	0.5814	0.7020	0.6576	0.7417	0.6621
730	0.4928	0.2550	0.0718	0.0814	0.1186	0.1036	0.0837	0.2024	0.3981	0.7701	0.5874	0.7059	0.6656	0.7491	0.6662
735	0.4960	0.2684	0.0731	0.0901	0.1243	0.1059	0.0864	0.2032	0.3990	0.7667	0.5885	0.7085	0.6641	0.7516	0.6726
740	0.4976	0.2862	0.0791	0.1042	0.1359	0.1123	0.0882	0.2074	0.4096	0.7735	0.5911	0.7047	0.6667	0.7532	0.6774
745	0.4993	0.3086	0.0828	0.1228	0.1466	0.1175	0.0923	0.2160	0.4187	0.7720	0.5878	0.7021	0.6688	0.7567	0.6834
750	0.5015	0.3262	0.0896	0.1482	0.1617	0.1217	0.0967	0.2194	0.4264	0.7739	0.5896	0.7071	0.6713	0.7600	0.6808
755	0.5044	0.3483	0.0980	0.1793	0.1739	0.1304	0.0996	0.2293	0.4370	0.7740	0.5947	0.7088	0.6657	0.7592	0.6838
760	0.5042	0.3665	0.1063	0.2129	0.1814	0.1330	0.1027	0.2378	0.4424	0.7699	0.5945	0.7055	0.6712	0.7605	0.6874
765	0.5073	0.3814	0.1137	0.2445	0.1907	0.1373	0.1080	0.2448	0.4512	0.7788	0.5935	0.7073	0.6745	0.7629	0.6955
770	0.5112	0.3974	0.1238	0.2674	0.1976	0.1376	0.1115	0.2489	0.4579	0.7801	0.5979	0.7114	0.6780	0.7646	0.7012
775	0.5147	0.4091	0.1381	0.2838	0.1958	0.1384	0.1118	0.2558	0.4596	0.7728	0.5941	0.7028	0.6744	0.7622	0.6996
780	0.5128	0.4206	0.1505	0.2979	0.1972	0.1390	0.1152	0.2635	0.4756	0.7793	0.5962	0.7105	0.6786	0.7680	0.7023
785	0.5108	0.4230	0.1685	0.3067	0.2018	0.1378	0.1201	0.2775	0.4880	0.7797	0.5919	0.7078	0.6823	0.7672	0.7022
790	0.5171	0.4397	0.1862	0.3226	0.2093	0.1501	0.1253	0.2957	0.5066	0.7754	0.5996	0.7112	0.6806	0.7645	0.7144
795	0.5135	0.4456	0.2078	0.3396	0.2161	0.1526	0.1313	0.3093	0.5214	0.7810	0.5953	0.7123	0.6718	0.7669	0.7062
800	0.5191	0.4537	0.2338	0.3512	0.2269	0.1646	0.1393	0.3239	0.5450	0.7789	0.5953	0.7158	0.6813	0.7683	0.7075
805	0.5191	0.4537	0.2338	0.3512	0.2269	0.1646	0.1393	0.3239	0.5450	0.7789	0.5953	0.7158	0.6813	0.7683	0.7075
810	0.5191	0.4537	0.2338	0.3512	0.2269	0.1646	0.1393	0.3239	0.5450	0.7789	0.5953	0.7158	0.6813	0.7683	0.7075
815	0.5191	0.4537	0.2338	0.3512	0.2269	0.1646	0.1393	0.3239	0.5450	0.7789	0.5953	0.7158	0.6813	0.7683	0.7075
820	0.5191	0.4537	0.2338	0.3512	0.2269	0.1646	0.1393	0.3239	0.5450	0.7789	0.5953	0.7158	0.6813	0.7683	0.7075
825	0.5191	0.4537	0.2338	0.3512	0.2269	0.1646	0.1393	0.3239	0.5450	0.7789	0.5953	0.7158	0.6813	0.7683	0.7075
830	0.5191	0.4537	0.2338	0.3512	0.2269	0.1646	0.1393	0.3239	0.5450	0.7789	0.5953	0.7158	0.6813	0.7683	0.7075

Appendix B
Matlab Codes for Colorimetric
and Photometric Calculations

Important note: All the codes are written for a spectral power distribution spanning 360–830 nm with a 5 nm interval.

© The Author(s), under exclusive licence to Springer Nature Singapore Pte Ltd. 2019
T. Erdem and H. V. Demir, *Color Science and Photometry for Lighting with LEDs and Semiconductor Nanocrystals*, Nanoscience and Nanotechnology,
https://doi.org/10.1007/978-981-13-5886-9

CHROMATICITY COORDINATE CALCULATION

```
function chromcalculator52 = chromcalculator52(spectrum)

%This function calculates the x,y chromaticity coordinates of a given
%spectral power distribution. spectrum is the array containing this
%information.

%colormatching() is the function having the color matcing functions
bars = colormatching();
xbar = bars(:,1);ybar = bars(:,2);zbar = bars(:,3);

X = sum(spectrum.*xbar);Y = sum(spectrum.*ybar);Z = sum(spectrum.*zbar);

x=X/(X+Y+Z);y=Y/(X+Y+Z);z=Z/(X+Y+Z);

chromcalculator52=[x y z Y];
```

CORRELATED COLOR TEMPERATURE CALCULATION

```
function newCCT=newCCT(x,y)

%this code calculates correlated color temperatures upto 20774 K
%(x,y) are the chromaticity coordinates

u=4*x/(-2*x+12*y+3);v=6*y/(-2*x+12*y+3);

%CCTarr is the list of (u,v) coordinates of the Planckian locus at
%different temperatures (the first column is the temperature in K, the
%second column contains u-coordinates and third column contains the v-
%coordinates in the range of 1000 K to 20774 K.

CCTarr=CCTarr();
mindistance=100000;
Tf=-1;
for i=2:length(CCTarr)-1

    ut=CCTarr(i,2);
    ute1=CCTarr(i-1,2);
    uta1=CCTarr(i+1,2);

    vt=CCTarr(i,3);
    vte1=CCTarr(i-1,3);
    vta1=CCTarr(i+1,3);

    distance=sqrt((ut-u)^2+(vt-v)^2);
    if distance<=mindistance
       mindistance=distance;
       if CCTarr(i,1)<=20774
          Tm=CCTarr(i,1);
          dm=distance;
```

```
            Tme1=CCTarr(i-1,1);
            dme1=sqrt((ute1-u)^2+(vte1-v)^2);

            Tma1=CCTarr(i+1,1);
            dma1=sqrt((uta1-u)^2+(vta1-v)^2);

            a=dme1/(Tme1-Tm)/(Tme1-Tma1);
            b=dm/(Tm-Tme1)/(Tm-Tma1);
            c=dma1/(Tma1-Tme1)/(Tma1-Tm);

            A=a+b+c;
            B=-(a*(Tma1+Tm)+b*(Tma1+Tme1)+c*(Tm+Tme1));
            C=a*Tm*Tma1+b*Tme1*Tma1+c*Tme1*Tm;
            T=-B/(2*A);

            duv=A*T^2+B*T+C;
            if v>vt
                signofduv=1;
            else
                signofduv=-1;
            end

            dT=Tma1-Tme1;

            dist=sqrt((uta1-ute1)^2+(vta1-vte1)^2);
            xdist=(dme1^2+dist^2-dma1^2)/(2*dist);

            Tsol=Tme1+xdist*(Tma1-Tme1)/dist;
            duvh=sqrt(dme1^2-xdist^2);
            linshift=Tsol+(T-Tsol)*duvh*1000;

            if duvh<0.001
                Tf=linshift;
            else
                Tf=T;
            end

        end
    end
end

newCCT=Tf;
```

LUMINOUS EFFICACY OF OPTICAL RADIATION CALCULATION

```
function LERcalc = LERcalc (spectrum)

%This code calculates the photopic luminous efficacy of optical %radiation
(LER)
peye=peye();

%calculate the product of eye sensitivity and the spectrum
sumprod = sum(peye.*spectrum);
%calculate the total optical power
Sumspec = sum(spectrum);

%calculate the luminous efficacy of optical radiation
```

```
LERcalc=683*sumprod/sumspec;
```

SCOTOPIC LUMINOUS EFFICACY OF OPTICAL RADIATION CALCULATION

```
function SLERcalc = SLERcalc (spectrum)

%This code calculates the scotopic luminous efficacy of optical radiation
%(SLER)
seye=seye();

%calculate the product of eye sensitivity and the spectrum
sumprod = sum(seye.*spectrum);
%calculate the total optical power
sumspec = sum(spectrum);

%calculate the scotopic luminous efficacy of optical radiation
LERcalc=1699*sumprod/sumspec;
```

MESOPIC LUMINANCE CALCULATION

```
function mesopic = mesopic(spectrum)
%This code calculates the mesopic luminance and takes the spectrum as input

m1 = 0.5;
m0 = 0;
k = sensitivities();
p = k(1,:); %photopic eye sensitivity function
s = k(2,:); %scotopic eye sensitivity function
Lp = 683*sum(spectrum.*p)*5; %photopic luminance
Ls = 1699*sum(spectrum.*s)*5; %scotopic luminance

a = 0.7670;
b = 0.3334;
Ls/Lp;
while abs(m0-m1)>=0.0001
    m0 = m1;
    Lmes2 = (m0*Lp+(1-m0)*Ls*683/1699)/(m0+(1-m0)*683/1699);
    m1 = a + b*log10(Lmes2);
end

if Lmes2 >= 5
    m = 1;
elseif Lmes2 <= 0.005
    m = 0;
else
    m = m0;
end

vmes = m*p + (1-m)*s;
vmes = vmes/max(vmes);
Lmes1 = 683/vmes(40)*sum(vmes.*wave)*5;
mesopic = Lmes1;
```

CRI CALCULATION

```
function criCalc5nm = criCalc5nm(wave)

%This code calculates the CRI of a given spectral power distribution.
%input array should be a row vector of size 95 (360nm:5nm:830nm)
```

```matlab
ri=test_color_samples();%reflectivities of test color samples
r1=ri(:,1);r2=ri(:,2);r3=ri(:,3);r4=ri(:,4);
r5=ri(:,5);r6=ri(:,6);r7=ri(:,7);r8=ri(:,8);

%scale test to make its Ytest=100
test=wave;
xctest=chromcalculator52(test/max(test));
mtest=test*100/xctest(4)/max(test);

%calcualte the appropriate reference source based on the color temperature
T=double(newCCT(xctest(1),xctest(2)));
if (T<=5000)
    c=3e8;
    k=1.3806504e-23;
    h=6.62606896e-34;
    %define the wavelength range
    w=360e-9:5e-9:830e-9;
    %calculate the reference planckian source
    ref=2*h*c^2./(w.^5.*(exp(h*c./(w*k*T))-1));
else
    %if the CCT is too high, the reference source should be one of the
    %standard D-daylight illuminants of CIE
    if (T>7000)
        s=daylight();
        x=-2.0064e9/T^3+1.9018e6/T^2+0.24748e3/T+0.23704;
        y=-3*x^2+2.87*x-0.275;
        m1=(-1.3515-1.7703*x+5.9114*y)/(0.0241+0.2562*x-0.7341*y);
        m2=(-0.300-31.4424*x+30.0717*y)/(0.0241+0.2562*x-0.7341*y);
        ref=(s(:,1)+m1*s(:,2)+m2*s(:,3))';
    else
        s=daylight();
        x=-4.6070e9/T^3+2.9678e6/T^2+0.09911e3/T+0.244063;
        y=-3*x^2+2.87*x-0.275;
        m1=(-1.3515-1.7703*x+5.9114*y)/(0.0241+0.2562*x-0.7341*y);
        m2=(-0.300-31.4424*x+30.0717*y)/(0.0241+0.2562*x-0.7341*y);
        ref=(s(:,1)+m1*s(:,2)+m2*s(:,3))';
    end
end

%scale ref to make its Yref=100
xcref=chromcalculator52(ref/max(ref));
mref=ref*100/xcref(4)/max(ref);

%calculate the chromaticity coordinates (u,v) for ref
cref=chromcalculator52(mref);
uref=4*cref(1)/(-2*cref(1)+12*cref(2)+3);
vref=6*cref(2)/(-2*cref(1)+12*cref(2)+3);
Yref=cref(4);
ref=mref;

%calculate (u,v) of test wave
ctest=chromcalculator52(mtest);
utest=4*ctest(1)/(-2*ctest(1)+12*ctest(2)+3);
vtest=6*ctest(2)/(-2*ctest(1)+12*ctest(2)+3);
Ytest=ctest(4);
test=mtest;

%calculate the color difference
dC=sqrt((uref-utest)^2+(vtest-vref)^2);
```

```
%calculate refi's
ref1=r1'.*ref;ref2=r2'.*ref;ref3=r3'.*ref;ref4=r4'.*ref;
ref5=r5'.*ref;ref6=r6'.*ref;ref7=r7'.*ref;ref8=r8'.*ref;

%calculate the test colors-> testi
test1=test.*r1';test2=test.*r2';test3=test.*r3';test4=test.*r4';
test5=test.*r5';test6=test.*r6';test7=test.*r7';test8=test.*r8';

%calculate the chromaticity coordinates (u,v) for refi's
cref1=chromcalculator52(ref1);
uref1=4*cref1(1)/(-2*cref1(1)+12*cref1(2)+3);
vref1=6*cref1(2)/(-2*cref1(1)+12*cref1(2)+3);
Yref1=cref1(4);

cref2=chromcalculator52(ref2);
uref2=4*cref2(1)/(-2*cref2(1)+12*cref2(2)+3);
vref2=6*cref2(2)/(-2*cref2(1)+12*cref2(2)+3);
Yref2=cref2(4);

cref3=chromcalculator52(ref3);
uref3=4*cref3(1)/(-2*cref3(1)+12*cref3(2)+3);
vref3=6*cref3(2)/(-2*cref3(1)+12*cref3(2)+3);
Yref3=cref3(4);

cref4=chromcalculator52(ref4);
uref4=4*cref4(1)/(-2*cref4(1)+12*cref4(2)+3);
vref4=6*cref4(2)/(-2*cref4(1)+12*cref4(2)+3);
Yref4=cref4(4);

cref5=chromcalculator52(ref5);
uref5=4*cref5(1)/(-2*cref5(1)+12*cref5(2)+3);
vref5=6*cref5(2)/(-2*cref5(1)+12*cref5(2)+3);
Yref5=cref5(4);

cref6=chromcalculator52(ref6);
uref6=4*cref6(1)/(-2*cref6(1)+12*cref6(2)+3);
vref6=6*cref6(2)/(-2*cref6(1)+12*cref6(2)+3);
Yref6=cref6(4);

cref7=chromcalculator52(ref7);
uref7=4*cref7(1)/(-2*cref7(1)+12*cref7(2)+3);
vref7=6*cref7(2)/(-2*cref7(1)+12*cref7(2)+3);
Yref7=cref7(4);

cref8=chromcalculator52(ref8);
uref8=4*cref8(1)/(-2*cref8(1)+12*cref8(2)+3);
vref8=6*cref8(2)/(-2*cref8(1)+12*cref8(2)+3);
Yref8=cref8(4);

%calculate the chromaticity coordinates (u,v) for testi
ctest1=chromcalculator52(test1);
utest1=4*ctest1(1)/(-2*ctest1(1)+12*ctest1(2)+3);
vtest1=6*ctest1(2)/(-2*ctest1(1)+12*ctest1(2)+3);
Ytest1=ctest1(4);

ctest2=chromcalculator52(test2);
utest2=4*ctest2(1)/(-2*ctest2(1)+12*ctest2(2)+3);
vtest2=6*ctest2(2)/(-2*ctest2(1)+12*ctest2(2)+3);
```

```
Ytest2=ctest2(4);

ctest3=chromcalculator52(test3);
utest3=4*ctest3(1)/(-2*ctest3(1)+12*ctest3(2)+3);
vtest3=6*ctest3(2)/(-2*ctest3(1)+12*ctest3(2)+3);
Ytest3=ctest3(4);

ctest4=chromcalculator52(test4);
utest4=4*ctest4(1)/(-2*ctest4(1)+12*ctest4(2)+3);
vtest4=6*ctest4(2)/(-2*ctest4(1)+12*ctest4(2)+3);
Ytest4=ctest4(4);

ctest5=chromcalculator52(test5);
utest5=4*ctest5(1)/(-2*ctest5(1)+12*ctest5(2)+3);
vtest5=6*ctest5(2)/(-2*ctest5(1)+12*ctest5(2)+3);
Ytest5=ctest5(4);

ctest6=chromcalculator52(test6);
utest6=4*ctest6(1)/(-2*ctest6(1)+12*ctest6(2)+3);
vtest6=6*ctest6(2)/(-2*ctest6(1)+12*ctest6(2)+3);
Ytest6=ctest6(4);

ctest7=chromcalculator52(test7);
utest7=4*ctest7(1)/(-2*ctest7(1)+12*ctest7(2)+3);
vtest7=6*ctest7(2)/(-2*ctest7(1)+12*ctest7(2)+3);
Ytest7=ctest7(4);

ctest8=chromcalculator52(test8);
utest8=4*ctest8(1)/(-2*ctest8(1)+12*ctest8(2)+3);
vtest8=6*ctest8(2)/(-2*ctest8(1)+12*ctest8(2)+3);
Ytest8=ctest8(4);

%calculate c's and d's of ref
refc=(4-uref-10*vref)/vref;
refd=(1.708*vref+0.404-1.481*uref)/vref;

%calculate c's and d's of testi
test1c=(4-utest1-10*vtest1)/vtest1;
test1d=(1.708*vtest1+0.404-1.481*utest1)/vtest1;

test2c=(4-utest2-10*vtest2)/vtest2;
test2d=(1.708*vtest2+0.404-1.481*utest2)/vtest2;

test3c=(4-utest3-10*vtest3)/vtest3;
test3d=(1.708*vtest3+0.404-1.481*utest3)/vtest3;

test4c=(4-utest4-10*vtest4)/vtest4;
test4d=(1.708*vtest4+0.404-1.481*utest4)/vtest4;

test5c=(4-utest5-10*vtest5)/vtest5;
test5d=(1.708*vtest5+0.404-1.481*utest5)/vtest5;

test6c=(4-utest6-10*vtest6)/vtest6;
test6d=(1.708*vtest6+0.404-1.481*utest6)/vtest6;

test7c=(4-utest7-10*vtest7)/vtest7;
test7d=(1.708*vtest7+0.404-1.481*utest7)/vtest7;
```

```
test8c=(4-utest8-10*vtest8)/vtest8;
test8d=(1.708*vtest8+0.404-1.481*utest8)/vtest8;

%calculate c's and d's of test wave
testc=(4-utest-10*vtest)/vtest;
testd=(1.708*vtest+0.404-1.481*utest)/vtest;

%calculate double asterix coordinates
uutest1=(10.872+0.404*refc/testc*test1c-
4*refd/testd*test1d)/(16.518+1.481*refc/testc*test1c-refd/testd*test1d);
vvtest1=5.520/(16.518+1.481*refc/testc*test1c-refd/testd*test1d);

uutest2=(10.872+0.404*refc/testc*test2c-
4*refd/testd*test2d)/(16.518+1.481*refc/testc*test2c-refd/testd*test2d);
vvtest2=5.520/(16.518+1.481*refc/testc*test2c-refd/testd*test2d);

uutest3=(10.872+0.404*refc/testc*test3c-
4*refd/testd*test3d)/(16.518+1.481*refc/testc*test3c-refd/testd*test3d);
vvtest3=5.520/(16.518+1.481*refc/testc*test3c-refd/testd*test3d);

uutest4=(10.872+0.404*refc/testc*test4c-
4*refd/testd*test4d)/(16.518+1.481*refc/testc*test4c-refd/testd*test4d);
vvtest4=5.520/(16.518+1.481*refc/testc*test4c-refd/testd*test4d);

uutest5=(10.872+0.404*refc/testc*test5c-
4*refd/testd*test5d)/(16.518+1.481*refc/testc*test5c-refd/testd*test5d);
vvtest5=5.520/(16.518+1.481*refc/testc*test5c-refd/testd*test5d);

uutest6=(10.872+0.404*refc/testc*test6c-
4*refd/testd*test6d)/(16.518+1.481*refc/testc*test6c-refd/testd*test6d);
vvtest6=5.520/(16.518+1.481*refc/testc*test6c-refd/testd*test6d);

uutest7=(10.872+0.404*refc/testc*test7c-
4*refd/testd*test7d)/(16.518+1.481*refc/testc*test7c-refd/testd*test7d);
vvtest7=5.520/(16.518+1.481*refc/testc*test7c-refd/testd*test7d);

uutest8=(10.872+0.404*refc/testc*test8c-
4*refd/testd*test8d)/(16.518+1.481*refc/testc*test8c-refd/testd*test8d);
vvtest8=5.520/(16.518+1.481*refc/testc*test8c-refd/testd*test8d);

uutest=(10.872+0.404*refc-4*refd)/(16.518+1.418*refc-refd);
vvtest=5.520/(16.518+1.418*refc-refd);

dl1=25*Yref1^(1/3)-25*Ytest1^(1/3);
dl2=25*Yref2^(1/3)-25*Ytest2^(1/3);
dl3=25*Yref3^(1/3)-25*Ytest3^(1/3);
dl4=25*Yref4^(1/3)-25*Ytest4^(1/3);
dl5=25*Yref5^(1/3)-25*Ytest5^(1/3);
dl6=25*Yref6^(1/3)-25*Ytest6^(1/3);
dl7=25*Yref7^(1/3)-25*Ytest7^(1/3);
dl8=25*Yref8^(1/3)-25*Ytest8^(1/3);

du1=13*(25*Yref1^(1/3)-17)*(uref1-uref)-13*(25*Ytest1^(1/3)-17)*(uutest1-
uutest);
du2=13*(25*Yref2^(1/3)-17)*(uref2-uref)-13*(25*Ytest2^(1/3)-17)*(uutest2-
uutest);
du3=13*(25*Yref3^(1/3)-17)*(uref3-uref)-13*(25*Ytest3^(1/3)-17)*(uutest3-
uutest);
```

```
du4=13*(25*Yref4^(1/3)-17)*(uref4-uref)-13*(25*Ytest4^(1/3)-17)*(uutest4-
uutest);
du5=13*(25*Yref5^(1/3)-17)*(uref5-uref)-13*(25*Ytest5^(1/3)-17)*(uutest5-
uutest);
du6=13*(25*Yref6^(1/3)-17)*(uref6-uref)-13*(25*Ytest6^(1/3)-17)*(uutest6-
uutest);
du7=13*(25*Yref7^(1/3)-17)*(uref7-uref)-13*(25*Ytest7^(1/3)-17)*(uutest7-
uutest);
du8=13*(25*Yref8^(1/3)-17)*(uref8-uref)-13*(25*Ytest8^(1/3)-17)*(uutest8-
uutest);

dv1=13*(25*Yref1^(1/3)-17)*(vref1-vref)-13*(25*Ytest1^(1/3)-17)*(vvtest1-
vvtest);
dv2=13*(25*Yref2^(1/3)-17)*(vref2-vref)-13*(25*Ytest2^(1/3)-17)*(vvtest2-
vvtest);
dv3=13*(25*Yref3^(1/3)-17)*(vref3-vref)-13*(25*Ytest3^(1/3)-17)*(vvtest3-
vvtest);
dv4=13*(25*Yref4^(1/3)-17)*(vref4-vref)-13*(25*Ytest4^(1/3)-17)*(vvtest4-
vvtest);
dv5=13*(25*Yref5^(1/3)-17)*(vref5-vref)-13*(25*Ytest5^(1/3)-17)*(vvtest5-
vvtest);
dv6=13*(25*Yref6^(1/3)-17)*(vref6-vref)-13*(25*Ytest6^(1/3)-17)*(vvtest6-
vvtest);
dv7=13*(25*Yref7^(1/3)-17)*(vref7-vref)-13*(25*Ytest7^(1/3)-17)*(vvtest7-
vvtest);
dv8=13*(25*Yref8^(1/3)-17)*(vref8-vref)-13*(25*Ytest8^(1/3)-17)*(vvtest8-
vvtest);

dE1=sqrt(dl1^2+du1^2+dv1^2);
dE2=sqrt(dl2^2+du2^2+dv2^2);
dE3=sqrt(dl3^2+du3^2+dv3^2);
dE4=sqrt(dl4^2+du4^2+dv4^2);
dE5=sqrt(dl5^2+du5^2+dv5^2);
dE6=sqrt(dl6^2+du6^2+dv6^2);
dE7=sqrt(dl7^2+du7^2+dv7^2);
dE8=sqrt(dl8^2+du8^2+dv8^2);

cri1=100-4.6*dE1;
cri2=100-4.6*dE2;
cri3=100-4.6*dE3;
cri4=100-4.6*dE4;
cri5=100-4.6*dE5;
cri6=100-4.6*dE6;
cri7=100-4.6*dE7;
cri8=100-4.6*dE8;

criCalc5nm=1/8*(cri1+cri2+cri3+cri4+cri5+cri6+cri7+cri8);
```

CQS CALCULATION

```
function cqsCalc=cqsCalc(wave)
%this code calculates color quality scale described by Yoshi Ohno
%input wavelength 360:5:830
wave=wave';

%ri vector will be filled with 15 Munsell color spectral reflection data
ri=munsell15();
%%%%CAUTION -TEMPERATURE!!!!
%creating the reference source
chrm=chromcalculator52(wave');
```

```matlab
T=newCCT(chrm(1),chrm(2));
if (T<=5000)
    %c=3e8;
    %k=1.3806504e-23;
    %h=6.62606896e-34;
    %define the wavelength range
    w=360e-9:5e-9:830e-9;
    %calculate the reference Planckian source
    i2=3.7415e-16;
    i3=0.014388;
    i1=T;
    ref=i2./(w.^5)./(exp(i3./(w*i1))-1);
    ref=ref*5;

else
    %if the CCT is too high, the reference source should be one of the
    %standard D-daylight illuminants of CIE
    if (T>7000)
        s=daylight();
        x=-2.0064e9/T^3+1.9018e6/T^2+0.24748e3/T+0.23704;
        y=-3*x^2+2.87*x-0.275;
        m1=(-1.3515-1.7703*x+5.9114*y)/(0.0241+0.2562*x-0.7341*y);
        m2=(0.030-31.4424*x+30.0717*y)/(0.0241+0.2562*x-0.7341*y);
        ref=(s(:,1)+m1*s(:,2)+m2*s(:,3))';
    else
        %this is the case for CCT>5000 K
        s=daylight();
        x=-4.6070e9/T^3+2.9678e6/T^2+0.09911e3/T+0.244063;
        y=-3*x^2+2.87*x-0.275;
        m1=(-1.3515-1.7703*x+5.9114*y)/(0.0241+0.2562*x-0.7341*y);
        m2=(0.030-31.4424*x+30.0717*y)/(0.0241+0.2562*x-0.7341*y);
        ref=(s(:,1)+m1*s(:,2)+m2*s(:,3))';
    end
end

%scale reference source to make its Yref=100
%ref'
xcref=chromcalculator52(ref);
Yref=xcref(4);
mref=100/Yref*ref;
ref=mref;
chrref=chromcalculator52(ref);

%calculate (L*,a*,b*) of reference source
LABref=LABcalc(chrref(1),chrref(2),chrref(4),chrref(1),chrref(2),chrref(4))
;

%scale test source to make its Ytest=100
test=wave;
xctest=chrm;
mtest=test*100/xctest(4);
test=mtest;
chrtest=chromcalculator52(test');

%calculate (L*,a*,b*) of test source
LABtest=LABcalc(chrtest(1),chrtest(2),chrtest(4),chrref(1),chrref(2),chrref
(4));

gamutArr=zeros(16,3);
% make the necessary calculations for every Munsell test samples
```

```
% refi stands for the reflectance spectrum of ith Munsell sample
illuminated with the reference source
% testi stands for the reflectance spectrum of ith Munsell sample
illuminated with the test source

% color data for D65 illuminant
% these data are going to be used in the CCT factor calculations
XD65=95.0474875793874;
YD65=100.0000000000000;
ZD65=108.9022607478380;
chrD65=[XD65/(XD65+YD65+ZD65) YD65/(XD65+YD65+ZD65) ZD65/(XD65+YD65+ZD65)
YD65];

for i=1:15
  r=ri(:,i);
  %calculate refi's (Munsell sample reflectance spectrum illuminated with
%the reference source)
  refi=r.*ref';
  %calculate testi's (Munsell sample reflectance spectrum illuminated
%with the test source)
  testi=test.*r;

  %make the CMCCAT2000 color adaptation for testi's
  chrtesti=chromcalculator52(testi');
  %input prameters for CMCCAT are:
xyzYorgwave,xyzYtestill,xyzYrefill,La1,La2
  chrtesti=CMCCAT2000(chrtesti,chrtest,chrref,1000,1000);
  %calculate the chromaticity coordinates (L*,a*,b*) for testi

LABtesti=LABcalc(chrtesti(1),chrtesti(2),chrtesti(4),chrref(1),chrref(2),ch
rref(4));

  chrrefi=chromcalculator52(refi');

  %calculate the chromaticity coordinates (L*,a*,b*) for refi's

LABrefi=LABcalc(chrrefi(1),chrrefi(2),chrrefi(4),chrref(1),chrref(2),chrref
(4));

  %calculate C*ab
  Cabrefi=sqrt(LABrefi(2)^2+LABrefi(3)^2);
  Cabtesti=sqrt(LABtesti(2)^2+LABtesti(3)^2);
  dCab=Cabtesti-Cabrefi;

  % refi stands for L* value of the sample illuminated with the reference
source
  % test stands for L* value of the sample illuminated with the test source
  dL=LABrefi(1)-LABtesti(1);
  da=LABrefi(2)-LABtesti(2);
  db=LABrefi(3)-LABtesti(3);
  dEe=sqrt(dL^2+da^2+db^2);
  % calculate the Euclidian color difference
  if dCab > 0
      dE(i)=sqrt(dEe^2-dCab^2);
  else
      dE(i)=dEe;
  end

  %make the calculations necessary for the gamut area calculations to find
CCT factor
```

```
    chrcmc=CMCCAT2000(chrrefi,chrref,chrD65,1000,1000);
    chrcmcref=CMCCAT2000(chrref,chrref,chrD65,1000,1000);

LABCMC=LABcalc(chrcmc(1),chrcmc(2),chrcmc(4),chrcmcref(1),chrcmcref(2),chrc
mcref(4));
    gamutArr(i,1)=LABCMC(1);
    gamutArr(i,2)=LABCMC(2);
    gamutArr(i,3)=LABCMC(3);

end

%calculate the root mean square of the color differences
dErms=sqrt(1/15*sum(dE.^2));

%calculate the gamut area of the reference light reflected from test
%samples
gamutArr(16,:)=gamutArr(1,:);
L=gamutArr(:,1);
a=gamutArr(:,2);
b=gamutArr(:,3);
for i=1:length(gamutArr)-1
    A(i)=sqrt(a(i)^2+b(i)^2);
    B(i)=sqrt(a(i+1)^2+b(i+1)^2);
    C(i)=sqrt((a(i+1)-a(i))^2+(b(i+1)-b(i))^2);
    t(i)=(A(i)+B(i)+C(i))/2;
    S(i)=sqrt(t(i)*(t(i)-A(i))*(t(i)-B(i))*(t(i)-C(i)));
end

area=sum(S);

if area > 8210
    cctFactor=1;
else
    cctFactor=area/8210;
end

alfa=3.105;
Rin=100-alfa*dErms;
%transform the scaling to [0,100] from [-100,100] and multiply by the
%cctFactor
cqsCalc=10*log(exp(Rin/10)+1)*cctFactor;
```

Lab COLOR COORDINATES CALCULATION

```
function LABcalc=LABcalc(xtest,ytest,Ytest,xref,yref,Yref)
%this code calculates L,a,b color  coordinates
factor=Ytest/ytest;X=factor*xtest;Y=factor*ytest;Z=factor*(1-xtest-ytest);

factor=Yref/yref;
Xn=factor*xref;Yn=factor*yref;Zn=factor*(1-xref-yref);

tx=X/Xn;ty=Y/Yn;tz=Z/Zn;

L=116*ty^(1/3)-16;a=500*(tx^(1/3)-ty^(1/3));b=200*(ty^(1/3)-tz^(1/3));

LABcalc=[L a b];
```

CMCCAT2000 CALCULATION

```
function CMCCAT2000=CMCCAT2000(xyzYorgwave,xyzYtestill,xyzYrefill,La1,La2)
%This code calculates CMCCAT2000 adaptation
%This calculation is based on Color Research and application, Vol. 27,
%Number 1, February 2002

%set F=1 for average illumination and F=0.8 for dim-and dark surround
F=1;

%calculate X,Y,Z of original wave
factor=xyzYorgwave(4)/xyzYorgwave(2);
X=xyzYorgwave(1)*factor;Y=xyzYorgwave(2)*factor;Z=xyzYorgwave(3)*factor;

%calculate X,Y,Z of test illuminant
factor=xyzYtestill(4)/xyzYtestill(2);
Xt=xyzYtestill(1)*factor;Yt=xyzYtestill(2)*factor;Zt=xyzYtestill(3)*factor;

%calculate X,Y,Z of referance illuminant
%make Y of refill to be 100
factor=xyzYrefill(4)/xyzYrefill(2);
Xr=xyzYrefill(1)*factor;Yr=xyzYrefill(2)*factor;Zr=xyzYrefill(3)*factor;

M=[0.7982  0.3389 -0.1371
  -0.5918  1.5512  0.0406
   0.0008  0.0239  0.9753];

%calculate the RGBs
RGB=M*[X Y Z]';%RGB of original wave
RGBt=M*[Xt Yt Zt]';%RGB of test illuminator
RGBr=M*[Xr Yr Zr]';%RGB of reference illuminator

%calculate the degree of adaptation
D=F*(0.08*log10(0.5*(La1+La2))+0.76-0.45*(La1-La2)/(La1+La2));
if D>1
   D=1;
elseif D<0
   D=0;
end

alfa=D;%*Yt/Yr;

%calculate adapted RGB
Rc=RGB(1)*(alfa*RGBr(1)/RGBt(1)+1-D);
Gc=RGB(2)*(alfa*RGBr(2)/RGBt(2)+1-D);
Bc=RGB(3)*(alfa*RGBr(3)/RGBt(3)+1-D);

invM=inv(M);
XYZc=invM*[Rc Gc Bc]';
s=sum(XYZc);
CMCCAT2000=[XYZc(1)/s XYZc(2)/s XYZc(3)/s XYZc(2)];
```

Printed in the United States
By Bookmasters